T0275479

Quantum Chemistry

Quantum Chemistry

Edited by
Ivor McGarry

Larsen & Keller
www.larsen-keller.com

Quantum Chemistry
Edited by Ivor McGarry
ISBN: 978-1-63549-244-6 (Hardback)

 Larsen & Keller

Published by Larsen and Keller Education,
5 Penn Plaza,
19th Floor,
New York, NY 10001, USA

Cataloging-in-Publication Data

Quantum chemistry / edited by Ivor McGarry.
   p. cm.
Includes bibliographical references and index.
ISBN 978-1-63549-244-6
1. Quantum chemistry. 2. Quantum theory. 3. Quantum electronics.
I. McGarry, Ivor.
QD462 .Q36 2017
541.28--dc23

The publisher's policy is to use permanent paper from mills that operate a sustainable forestry policy. Furthermore, the publisher ensures that the text paper and cover boards used have met acceptable environmental accreditation standards.

Printed and bound in the United States of America.

For more information regarding Larsen and Keller Education and its products, please visit the publisher's website www.larsen-keller.com

Table of Contents

Preface   **VII**

Chapter 1   **Introduction to Quantum Chemistry**   **1**

Chapter 2   **Key Concepts of Quantum Chemistry**   **6**
     i. Electron Configuration   6
     ii. Molecular Symmetry   16
     iii. Interatomic Potential   28
     iv. Quantum Harmonic Oscillator   33
     v. Franck–Condon Principle   46
     vi. Born–Oppenheimer Approximation   54
     vii. Jellium   61
     viii. Quantum Chemistry Composite Methods   65

Chapter 3   **Electronic Structure and Theories**   **72**
     i. VSEPR Theory   72
     ii. Valence Bond Theory   82
     iii. Molecular Orbital Theory   85
     iv. Frontier Molecular Orbital Theory   89
     v. Ligand Field Theory   93
     vi. Crystal Field Theory   96
     vii. Bohr Model   102
     viii. Old Quantum Theory   115

Chapter 4   **Orbitals and Hydrogen Atoms**   **127**
     i. Atomic Orbital   127
     ii. Molecular Orbital   146
     iii. Molecular Orbital Diagram   156
     iv. Hydrogen Atom   167
     v. Hydrogen-like Atom   175

Chapter 5   **Perturbation Theory and Variational Method**   **186**
     i. Variational Method (Quantum Mechanics)   186
     ii. Perturbation Theory   190
     iii. Singular Perturbation   201
     iv. Ab Initio Quantum Chemistry Methods   203

Chapter 6   **Semi-Empirical Quantum Chemistry Method**   **209**
     i. Semi-empirical Quantum Chemistry Method   209
     ii. Extended Hückel Method   210
     iii. PM3 (Chemistry)   212
     iv. Zero Differential Overlap   213

Chapter 7   **Understanding Diatomic Molecule**   **215**
    i.   Diatomic Molecule   215
    ii.   Hund's Cases   220
    iii. Symmetry of Diatomic Molecules   222
    iv.   Molecular Geometry   239

Chapter 8   **Density Functional Theory**   **247**
    i.   Electronic Density   247
    ii.   Density Functional Theory   249
    iii. Time-dependent Density Functional Theory   258
    iv.   Lieb–Oxford Inequality   262
    v.   Minnesota Functionals   266
    vi.   Runge–Gross Theorem   268
    vii. Orbital-free Density Functional Theory   271
    viii. Local-density Approximation   272

Permissions

Index

Preface

Quantum chemistry is the study of chemical properties of matter along with the application of theorems and methods derived from quantum mechanics. The field is relatively new and it concentrates on the states of the nucleus as well as the molecule; hence it is applied in computational chemistry and atomic physics. This textbook on quantum chemistry discusses the fundamental concepts in the most easy and comprehensive manner. The field is highly experimental and the topics included elucidate the various technology and devices that are used to study this field. Different approaches, evaluations and methodologies on quantum chemistry have been included in this book. The textbook aims to serve as a resource guide for students and facilitate the study of the subject.

A short introduction to every chapter is written below to provide an overview of the content of the book:

Chapter 1 - Quantum chemistry studies the application of the theories and principles of quantum mechanics in experiments and models of chemical structures. Quantum chemistry concerns itself with experimental and theoretical methods. This chapter is an overview of the subject matter incorporating all the major aspects of quantum chemistry; **Chapter 2** - To understand chemical bonding and molecular behavior it is imperative to fathom the atomic structure and the electron configuration. The content of this chapter is intended to cater to a better understanding of the key concepts of quantum chemistry such as molecular symmetry, interatomic potential, electron configuration, quantum harmonic oscillator, Frank-Condon principle, Born-Oppenheimer approximation and jellium. These principles, concepts and models form an integral part of quantum chemistry as they outline the structure of atomic nuclei and the arrangement of electrons; **Chapter 3** - Chemical bonding is related to the sharing or transference of electrons between atoms and in an attempt to grasp the atomic workings in the bonding of atoms, numerous models and theories have been developed. This chapter illustrates models like VSEPR theory, valence bond theory, molecular orbital theory, frontier molecular orbital theory, Ligand field theory, crystal field theory, Bohr model and old quantum theory; **Chapter 4** - Atomic orbitals are the areas of the atom where the probability of finding electrons is high and these orbitals are designated by three exclusive quantum numbers n, l and m. Molecular orbital on the other hand describes the chemical or physical properties of an electron in a molecule. In the chapter the reader is introduced to molecular and atomic orbitals and the molecular orbital theory. There is a section on the hydrogen atom and systems like muonium, positronium etc. that can be categorized as hydrogen-like atoms; **Chapter 5** - The Schrödinger equation in quantum chemistry is solved by approximate methods and this chapter discusses the two primary methods- perturbation theory and the variational method. The perturbation theory uses the reference of an unperturbed

Hamiltonian to approximately explain the perturbation. But when it becomes difficult to pin down an unperturbed reference Hamiltonian, the variational method employs the use of a trial wavefunction. This chapter illustrates these methods in-depth; **Chapter 6** – Semi-empirical quantum chemistry Method is an important concept in computational chemistry. Along with semi-empirical quantum chemistry, this chapter also focuses on extended Hückel method, PM3 and zero differential overlap. The section strategically encompasses and incorporates the major components of semi-empirical quantum chemistry method, providing a complete understanding; **Chapter 7** – The molecules formed by only two atoms are referred to as diatomic molecules and diatomic molecules can be composed of the same atom in case of oxygen (O2) or of two different molecules like nitric oxide (NO). This chapter explores the subject of diatomic molecules by imputing to Hund's cases and the symmetry of diatomic molecules; **Chapter 8** – To determine the solution of the Schrödinger equation for an N-body system, the density functional theory is employed. Electron density is used as the functional or the underlying property and the electron density functional in turn helps determine the total energy of the system. This chapter explores the density functional theory, its contributing factors like Lieb–Oxford inequality, Local-density approximation and the Minnesota functionals; the Runge–Gross theorem which forms a basis for the time-dependent density functional theory and the orbital-free density functional theory.

I extend my sincere thanks to the publisher for considering me worthy of this task. Finally, I thank my family for being a source of support and help.

Editor

Introduction to Quantum Chemistry

Quantum chemistry studies the application of the theories and principles of quantum mechanics in experiments and models of chemical structures. Quantum chemistry concerns itself with experimental and theoretical methods. This chapter is an overview of the subject matter incorporating all the major aspects of quantum chemistry.

Quantum Chemistry

Quantum chemistry is a branch of chemistry whose primary focus is the application of quantum mechanics in physical models and experiments of chemical systems. It is also called molecular quantum mechanics.

Overview

It involves heavy interplay of experimental and theoretical methods:

- Experimental quantum chemists rely heavily on spectroscopy, through which information regarding the quantization of energy on a molecular scale can be obtained. Common methods are infra-red (IR) spectroscopy and nuclear magnetic resonance (NMR) spectroscopy.

- Theoretical quantum chemistry, the workings of which also tend to fall under the category of computational chemistry, seeks to calculate the predictions of quantum theory as atoms and molecules can only have discrete energies; as this task, when applied to polyatomic species, invokes the many-body problem, these calculations are performed using computers rather than by analytical "back of the envelope" methods, pen recorder or computerized data station with a VDU.

In these ways, quantum chemists investigate chemical phenomena.

- In reactions, quantum chemistry studies the ground state of individual atoms and molecules, the excited states, and the transition states that occur during chemical reactions.

- On the calculations: quantum chemical studies use also semi-empirical and other methods based on quantum mechanical principles, and deal with time dependent problems. Many quantum chemical studies assume the nuclei are at rest (Born–Oppenheimer approximation). Many calculations involve iterative methods that include self-consistent field methods. Major goals of quantum chemistry include increasing the accuracy of the results for small molecular systems, and increasing the size of large molecules that can be processed, which is limited by scaling considerations—the computation time increases as a power of the number of atoms.

History

Some view the birth of quantum chemistry in the discovery of the Schrödinger equation and its application to the hydrogen atom in 1926. However, the 1927 article of Walter Heitler and Fritz London is often recognised as the first milestone in the history of quantum chemistry. This is the first application of quantum mechanics to the diatomic hydrogen molecule, and thus to the phenomenon of the chemical bond. In the following years much progress was accomplished by Edward Teller, Robert S. Mulliken, Max Born, J. Robert Oppenheimer, Linus Pauling, Erich Hückel, Douglas Hartree, Vladimir Aleksandrovich Fock, to cite a few. The history of quantum chemistry also goes through the 1838 discovery of cathode rays by Michael Faraday, the 1859 statement of the black body radiation problem by Gustav Kirchhoff, the 1877 suggestion by Ludwig Boltzmann that the energy states of a physical system could be discrete, and the 1900 quantum hypothesis by Max Planck that any energy radiating atomic system can theoretically be divided into a number of discrete energy elements ε such that each of these energy elements is proportional to the frequency v with which they each individually radiate energy and a numerical value called Planck's Constant. Then, in 1905, to explain the photoelectric effect (1839), i.e., that shining light on certain materials can function to eject electrons from the material, Albert Einstein postulated, based on Planck's quantum hypothesis, that light itself consists of individual quantum particles, which later came to be called photons (1926). In the years to follow, this theoretical basis slowly began to be applied to chemical structure, reactivity, and bonding. Probably the greatest contribution to the field was made by Linus Pauling.

Electronic Structure

The first step in solving a quantum chemical problem is usually solving the Schrödinger equation (or Dirac equation in relativistic quantum chemistry) with the electronic molecular Hamiltonian. This is called determining the electronic structure of the molecule. It can be said that the electronic structure of a molecule or crystal implies essentially its chemical properties. An exact solution for the Schrödinger equation can only be obtained for the hydrogen atom (though exact solutions for the bound state energies of the hydrogen molecular ion have been identified in terms of the generalized Lambert W function). Since all other atomic, or molecular systems, involve the motions of three or more "particles", their Schrödinger equations cannot be solved exactly and so approximate solutions must be sought.

Wave Model

The foundation of quantum mechanics and quantum chemistry is the wave model, in which the atom is a small, dense, positively charged nucleus surrounded by electrons. The wave model is derived from the wavefunction, a set of possible equations derived from the time evolution of the Schrödinger equation which is applied to the wavelike probability distribution of subatomic particles. Unlike the earlier Bohr model of the atom, however, the wave model describes electrons as "clouds" moving in orbitals, and their positions are represented by probability distributions rather than discrete points. The strength of this model lies in its predictive power. Specifically, it predicts the pattern of chemically similar elements found in the periodic table. The wave model is so named because electrons exhibit properties (such as interference) traditionally associated with waves. In this model, when we solve the Schrödinger Equation for an Hidrogenoid Atom, we obtain a solution that depends on some numbers, called quantum numbers, that describes the orbital, the most probable space where an elec-tron can be. These are n, the principal quantum number, for the energy, l, or secondary quantum number, which correlates to the angular momentum, ml, for the orientation, and ms the spin. This model can explain the new lines that appeared in the spectroscopy of atoms. For multielectron atoms we must introduce some rules as that the electrons fill orbitals in a way to minimize the energy of the atom, in order of increasing energy, the Pauli Exclusion Principle, the Hund's Rule, and the Aufbau Principle.

Valence Bond

Although the mathematical basis of quantum chemistry had been laid by Schrödinger in 1926, it is generally accepted that the first true calculation in quantum chemistry was that of the German physicists Walter Heitler and Fritz London on the hydrogen (H_2) molecule in 1927. Heitler and London's method was extended by the American theoretical physicist John C. Slater and the American theoretical chemist Linus Pauling to become the Valence-Bond (VB) [or Heitler–London–Slater–Pauling (HLSP)] meth-od. In this method, attention is primarily devoted to the pairwise interactions between atoms, and this method therefore correlates closely with classical chemists' drawings of bonds. It focuses on how the atomic orbitals of an atom combine to give individual chemical bonds when a molecule is formed. The concept of chemical bond disorts when the aromatic compounds are considered, then you need to apply resonance ideas and hybridization that doesn´t correspond to chemical view of fixed shared pair of elec-trons between molecules.

Molecular Orbital

An alternative approach was developed in 1929 by Friedrich Hund and Robert S. Mul-liken, in which electrons are described by mathematical functions delocalized over an entire molecule. The Hund–Mulliken approach or molecular orbital (MO) method is less intuitive to chemists, but has turned out capable of predicting spectroscopic properties better than the VB method. This approach is the conceptional basis of the Hartree–Fock method and further post Hartree–Fock methods.

Density Functional Theory

The Thomas–Fermi model was developed independently by Thomas and Fermi in 1927. This was the first attempt to describe many-electron systems on the basis of electronic density instead of wave functions, although it was not very successful in the treatment of entire molecules. The method did provide the basis for what is now known as density functional theory. Modern day DFT uses the Kohn-Sham method, where the density functional is split into four terms; the Kohn-Sham kinetic energy, an external potential, exchange and correlation energies. A large part of the focus on developing DFT is on improving the exchange and correlation terms. Though this method is less developed than post Hartree–Fock methods, its significantly lower computational requirements (scaling typically no worse than n^3 with respect to n basis functions, for the pure functionals) allow it to tackle larger polyatomic molecules and even macromolecules. This computational affordability and often comparable accuracy to MP2 and CCSD(T) (post-Hartree–Fock methods) has made it one of the most popular methods in computational chemistry at present.

Chemical Dynamics

A further step can consist of solving the Schrödinger equation with the total molecular Hamiltonian in order to study the motion of molecules. Direct solution of the Schrödinger equation is called *quantum molecular dynamics*, within the semiclassical approximation *semiclassical molecular dynamics*, and within the classical mechanics framework *molecular dynamics (MD)*. Statistical approaches, using for example Monte Carlo methods, are also possible.

Adiabatic Chemical Dynamics

In adiabatic dynamics, interatomic interactions are represented by single scalar potentials called potential energy surfaces. This is the Born–Oppenheimer approximation introduced by Born and Oppenheimer in 1927. Pioneering applications of this in chemistry were performed by Rice and Ramsperger in 1927 and Kassel in 1928, and generalized into the RRKM theory in 1952 by Marcus who took the transition state theory developed by Eyring in 1935 into account. These methods enable simple estimates of unimolecular reaction rates from a few characteristics of the potential surface.

Non-adiabatic Chemical Dynamics

Non-adiabatic dynamics consists of taking the interaction between several coupled potential energy surface (corresponding to different electronic quantum states of the molecule). The coupling terms are called vibronic couplings. The pioneering work in this field was done by Stueckelberg, Landau, and Zener in the 1930s, in their work on what is now known as the Landau–Zener transition. Their formula allows the transition probability between two diabatic potential curves in the neighborhood of an avoided crossing to be calculated.

References

- Atkins, P.W.; Friedman, R. (2005). Molecular Quantum Mechanics (4th ed.). Oxford University Press. ISBN 978-0-19-927498-7.

- Pullman, Bernard; Pullman, Alberte (1963). Quantum Biochemistry. New York and London: Academic Press. ISBN 90-277-1830-X.

- Kostas Gavroglu, Ana Simões: NEITHER PHYSICS NOR CHEMISTRY.A History of Quantum Chemistry, MIT Press, 2011, ISBN 0-262-01618-4

- Szabo, Attila; Ostlund, Neil S. (1996). Modern Quantum Chemistry: Introduction to Advanced Electronic Structure Theory. Dover. ISBN 0-486-69186-1.

- Landau, L.D.; Lifshitz, E.M. Quantum Mechanics:Non-relativistic Theory. Course of Theoretical Physic. 3. Pergamon Press. ISBN 0-08-019012-X.

- Pauling, L.; Wilson, E. B. (1963) [1935]. Introduction to Quantum Mechanics with Applications to Chemistry. Dover Publications. ISBN 0-486-64871-0.

- Simon, Z. (1976). Quantum Biochemistry and Specific Interactions. Taylor & Francis. ISBN 978-0-85626-087-2.

Key Concepts of Quantum Chemistry

To understand chemical bonding and molecular behavior it is imperative to fathom the atomic structure and the electron configuration. The content of this chapter is intended to cater to a better understanding of the key concepts of quantum chemistry such as molecular symmetry, interatomic potential, electron configuration, quantum harmonic oscillator, Frank-Condon principle, Born-Oppenheimer approximation and jellium. These principles, concepts and models form an integral part of quantum chemistry as they outline the structure of atomic nuclei and the arrangement of electrons.

Electron Configuration

In atomic physics and quantum chemistry, the electron configuration is the distribution of electrons of an atom or molecule (or other physical structure) in atomic or molecular orbitals. For example, the electron configuration of the neon atom is $1s^2\ 2s^2\ 2p^6$.

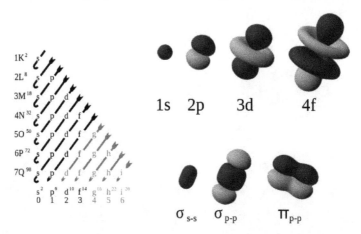

$$1s^2_2\,2s^2_4\,2p^6_{10}\,3s^2_{12}\,3p^6_{18}\,4s^2_{20}\,3d^{10}_{30}\,4p^6_{36}\,5s^2_{38}\,4d^{10}_{48}\,5p^6_{54}\,6s^2_{56}\,4f^{14}_{70}\,5d^{10}_{80}\,6p^6_{86}\,7s^2_{88}\,5f^{14}_{102}\,6d^{10}_{112}\,7p^6_{118}$$

Electron atomic and molecular orbitals

Electronic configurations describe electrons as each moving independently in an orbital, in an average field created by all other orbitals. Mathematically, configurations are described by Slater determinants or configuration state functions.

According to the laws of quantum mechanics, for systems with only one electron, an energy is associated with each electron configuration and, upon certain conditions, electrons are able to move from one configuration to another by the emission or absorption of a quantum of energy, in the form of a photon.

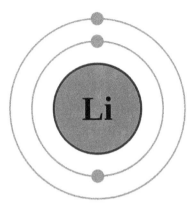

A Bohr diagram of lithium

Knowledge of the electron configuration of different atoms is useful in understanding the structure of the periodic table of elements. The concept is also useful for describing the chemical bonds that hold atoms together. In bulk materials, this same idea helps explain the peculiar properties of lasers and semiconductors.

Shells and Subshells

	s (ℓ=0)	p (ℓ=1)		
	m=0	m=0	m=±1	
	s	p_z	p_x	p_y
n=1	•			
n=2	•	⬤	⬤	⬤

Electron configuration was first conceived of under the Bohr model of the atom, and it is still common to speak of shells and subshells despite the advances in understanding of the quantum-mechanical nature of electrons.

An electron shell is the set of allowed states that share the same principal quantum number, n (the number before the letter in the orbital label), that electrons may occupy. An atom's nth electron shell can accommodate $2n^2$ electrons, e.g. the first shell can accommodate 2 electrons, the second shell 8 electrons, and the third shell 18 electrons. The factor of two arises because the allowed states are doubled due to electron spin—each atomic orbital admits up to two otherwise identical electrons with opposite spin, one with a spin +1/2 (usually denoted by an up-arrow) and one with a spin −1/2 (with a down-arrow).

A subshell is the set of states defined by a common azimuthal quantum number, ℓ, within a shell. The values ℓ = 0, 1, 2, 3 correspond to the s, p, d, and f labels, respectively. The maximum number of electrons that can be placed in a subshell is given by $2(2\ell + 1)$. This gives two electrons in an s subshell, six electrons in a p subshell, ten electrons in a d subshell and fourteen electrons in an f subshell.

The numbers of electrons that can occupy each shell and each subshell arise from the equations of quantum mechanics, in particular the Pauli exclusion principle, which states that no two electrons in the same atom can have the same values of the four quantum numbers.

Notation

Physicists and chemists use a standard notation to indicate the electron configurations of atoms and molecules. For atoms, the notation consists of a sequence of atomic orbital labels (e.g. for phosphorus the sequence 1s, 2s, 2p, 3s, 3p) with the number of electrons assigned to each orbital (or set of orbitals sharing the same label) placed as a superscript. For example, hydrogen has one electron in the s-orbital of the first shell, so its configuration is written $1s^1$. Lithium has two electrons in the 1s-subshell and one in the (higher-energy) 2s-subshell, so its configuration is written $1s^2\ 2s^1$ (pronounced "one-s-two, two-s-one"). Phosphorus (atomic number 15) is as follows: $1s^2\ 2s^2\ 2p^6\ 3s^2\ 3p^3$.

For atoms with many electrons, this notation can become lengthy and so an abbreviated notation is used, since all but the last few subshells are identical to those of one or another of the noble gases. Phosphorus, for instance, differs from neon ($1s^2\ 2s^2\ 2p^6$) only by the presence of a third shell. Thus, the electron configuration of neon is pulled out, and phosphorus is written as follows: $[Ne]\ 3s^2\ 3p^3$. This convention is useful as it is the electrons in the outermost shell that most determine the chemistry of the element.

For a given configuration, the order of writing the orbitals is not completely fixed since only the orbital occupancies have physical significance. For example, the electron configuration of the titanium ground state can be written as either $[Ar]\ 4s^2\ 3d^2$ or $[Ar]\ 3d^2\ 4s^2$. The first notation follows the order based on the Madelung rule for the configurations of neutral atoms; 4s is filled before 3d in the sequence Ar, K, Ca, Sc, Ti. The second notation groups all orbitals with the same value of n together, corresponding to the "spectroscopic" order of orbital energies that is the reverse of the order in which electrons are removed from a given atom to form positive ions; 3d is filled before 4s in the sequence Ti^{4+}, Ti^{3+}, Ti^{2+}, Ti^+, Ti.

The superscript 1 for a singly occupied orbital is not compulsory. It is quite common to see the letters of the orbital labels (s, p, d, f) written in an italic or slanting typeface, although the International Union of Pure and Applied Chemistry (IUPAC) recommends a normal typeface (as used here). The choice of letters originates from a now-obsolete system of categorizing spectral lines as "sharp", "principal", "diffuse" and "fundamental" (or "fine"), based on their observed fine structure: their modern usage indicates orbitals with an azimuthal quantum number, l, of 0, 1, 2 or 3 respectively. After "f", the sequence continues alphabetically "g", "h", "i"... (l = 4, 5, 6...), skipping «j», although orbitals of these types are rarely required.

The electron configurations of molecules are written in a similar way, except that mo-lecular orbital labels are used instead of atomic orbital labels .

Energy — Ground State and Excited States

The energy associated to an electron is that of its orbital. The energy of a configuration is often approximated as the sum of the energy of each electron, neglecting the elec-tron-electron interactions. The configuration that corresponds to the lowest electronic energy is called the ground state. Any other configuration is an excited state.

As an example, the ground state configuration of the sodium atom is $1s^2 2s^2 2p^6 3s$, as deduced from the Aufbau principle. The first excited state is obtained by promoting a 3s electron to the 3p orbital, to obtain the $1s^2 2s^2 2p^6 3p$ configuration, ab-breviated as the 3p level. Atoms can move from one configuration to another by absorb-ing or emitting energy. In a sodium-vapor lamp for example, sodium atoms are excited to the 3p level by an electrical discharge, and return to the ground state by emitting yellow light of wavelength 589 nm.

Usually, the excitation of valence electrons (such as 3s for sodium) involves energies corresponding to photons of visible or ultraviolet light. The excitation of core electrons is possible, but requires much higher energies, generally corresponding to x-ray pho-tons. This would be the case for example to excite a 2p electron to the 3s level and form the excited $1s^2 2s^2 2p^5 3s^2$ configuration.

The remainder of this article deals only with the ground-state configuration, often re-ferred to as "the" configuration of an atom or molecule.

History

Niels Bohr (1923) was the first to propose that the periodicity in the properties of the el-ements might be explained by the electronic structure of the atom. His proposals were based on the then current Bohr model of the atom, in which the electron shells were orbits at a fixed distance from the nucleus. Bohr's original configurations would seem strange to a present-day chemist: sulfur was given as 2.4.4.6 instead of $1s^2 2s^2 2p^6 3s^2 3p^4$ (2.8.6).

The following year, E. C. Stoner incorporated Sommerfeld's third quantum number into the description of electron shells, and correctly predicted the shell structure of sulfur to be 2.8.6. However neither Bohr's system nor Stoner's could correctly describe the changes in atomic spectra in a magnetic field (the Zeeman effect).

Bohr was well aware of this shortcoming (and others), and had written to his friend

Wolfgang Pauli to ask for his help in saving quantum theory (the system now known as "old quantum theory"). Pauli realized that the Zeeman effect must be due only to the outermost electrons of the atom, and was able to reproduce Stoner's shell structure, but with the correct structure of subshells, by his inclusion of a fourth quantum number and his exclusion principle (1925):

It should be forbidden for more than one electron with the same value of the main quantum number n *to have the same value for the other three quantum numbers* k [l], j [m_j] *and* m [m_s].

The Schrödinger equation, published in 1926, gave three of the four quantum numbers as a direct consequence of its solution for the hydrogen atom: this solution yields the atomic orbitals that are shown today in textbooks of chemistry (and above). The examination of atomic spectra allowed the electron configurations of atoms to be determined experimentally, and led to an empirical rule for the order in which atomic orbitals are filled with electrons.

Atoms: Aufbau Principle and Madelung Rule

The Aufbau principle (from the German *Aufbau*, "building up, construction") was an important part of Bohr's original concept of electron configuration. It may be stated as:

> *a maximum of two electrons are put into orbitals in the order of increasing orbital energy: the lowest-energy orbitals are filled before electrons are placed in higher-energy orbitals.*

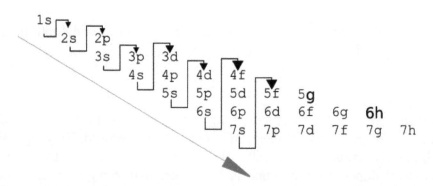

The approximate order of filling of atomic orbitals, following the arrows from 1s to 7p. (After 7p the order includes orbitals outside the range of the diagram, starting with 8s.)

The principle works very well (for the ground states of the atoms) for the first 18 elements, then decreasingly well for the following 100 elements. The modern form of the Aufbau principle describes an order of orbital energies given by Madelung's rule (or Klechkowski's rule). This rule was first stated by Charles Janet in 1929, rediscovered by Erwin Madelung in 1936, and later given a theoretical justification by V.M. Klechkowski

1.   Orbitals are filled in the order of increasing $n+l$;

2.   Where two orbitals have the same value of $n+l$, they are filled in order of increasing n.

This gives the following order for filling the orbitals:

1s, 2s, 2p, 3s, 3p, 4s, 3d, 4p, 5s, 4d, 5p, 6s, 4f, 5d, 6p, 7s, 5f, 6d, 7p, (8s, 5g, 6f, 7d, 8p, and 9s)

In this list the orbitals in parentheses are not occupied in the ground state of the heaviest atom now known (Uuo, Z = 118).

The Aufbau principle can be applied, in a modified form, to the protons and neutrons in the atomic nucleus, as in the shell model of nuclear physics and nuclear chemistry.

Periodic Table

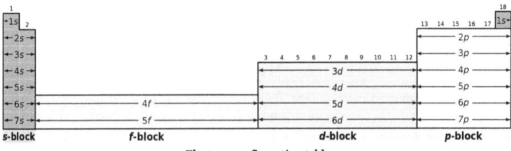

Electron configuration table

The form of the periodic table is closely related to the electron configuration of the atoms of the elements. For example, all the elements of group 2 have an electron configuration of [E] ns^2 (where [E] is an inert gas configuration), and have notable similarities in their chemical properties. In general, the periodicity of the periodic table in terms of periodic table blocks is clearly due to the number of electrons (2, 6, 10, 14...) needed to fill s, p, d, and f subshells.

The outermost electron shell is often referred to as the "valence shell" and (to a first approximation) determines the chemical properties. It should be remembered that the similarities in the chemical properties were remarked on more than a century before the idea of electron configuration. It is not clear how far Madelung's rule *explains* (rather than simply describes) the periodic table, although some properties (such as the common +2 oxidation state in the first row of the transition metals) would obviously be different with a different order of orbital filling.

Shortcomings of The Aufbau Principle

The Aufbau principle rests on a fundamental postulate that the order of orbital energies is fixed, both for a given element and between different elements; in both cases this is only approximately true. It considers atomic orbitals as "boxes" of fixed energy into which can be placed two electrons and no more. However, the energy of an electron "in" an atomic orbital depends on the energies of all the other electrons of the atom (or ion, or molecule, etc.). There are no "one-electron solutions" for systems of more than one electron, only a set of many-electron solutions that cannot be calculated exactly (although there are mathematical approximations available, such as the Hartree–Fock method).

The fact that the Aufbau principle is based on an approximation can be seen from the fact that there is an almost-fixed filling order at all, that, within a given shell, the s-orbital is always filled before the p-orbitals. In a hydrogen-like atom, which only has one electron, the s-orbital and the p-orbitals of the same shell have exactly the same energy, to a very good approximation in the absence of external electromagnetic fields. (However, in a real hydrogen atom, the energy levels are slightly split by the magnetic field of the nucleus, and by the quantum electrodynamic effects of the Lamb shift.)

Ionization of the Transition Metals

The naïve application of the Aufbau principle leads to a well-known paradox (or apparent paradox) in the basic chemistry of the transition metals. Potassium and calcium appear in the periodic table before the transition metals, and have electron configurations [Ar] $4s^1$ and [Ar] $4s^2$ respectively, i.e. the 4s-orbital is filled before the 3d-orbital. This is in line with Madelung's rule, as the 4s-orbital has $n+l = 4$ ($n = 4, l = 0$) while the 3d-orbital has $n+l = 5$ ($n = 3, l = 2$). After calcium, most neutral atoms in the first series of transition metals (Sc-Zn) have configurations with two 4s electrons, but there are two exceptions. Chromium and copper have electron configurations [Ar] $3d^5 4s^1$ and [Ar] $3d^{10} 4s^1$ respectively, i.e. one electron has passed from the 4s-orbital to a 3d-orbital to generate a half-filled or filled subshell. In this case, the usual explanation is that "half-filled or completely filled subshells are particularly stable arrangements of electrons".

The apparent paradox arises when electrons are *removed* from the transition metal atoms to form ions. The first electrons to be ionized come not from the 3d-orbital, as one would expect if it were "higher in energy", but from the 4s-orbital. This interchange of electrons between 4s and 3d is found for all atoms of the first series of transition metals. The configurations of the neutral atoms (K, Ca, Sc, Ti, V, Cr, ...) usually follow the order 1s, 2s, 2p, 3s, 3p, 4s, 3d, ...; however the successive stages of ionization of a given atom (such as Fe^{4+}, Fe^{3+}, Fe^{2+}, Fe^+, Fe) usually follow the order 1s, 2s, 2p, 3s, 3p, 3d, 4s, ...

This phenomenon is only paradoxical if it is assumed that the energy order of atomic orbitals is fixed and unaffected by the nuclear charge or by the presence of electrons in other orbitals. If that were the case, the 3d-orbital would have the same energy as the 3p-orbital, as it does in hydrogen, yet it clearly doesn't. There is no special reason why the Fe^{2+} ion should have the same electron configuration as the chromium atom, given that iron has two more protons in its nucleus than chromium, and that the chemistry of the two species is very different. Melrose and Eric Scerri have analyzed the changes of orbital energy with orbital occupations in terms of the two-electron repulsion integrals of the Hartree-Fock method of atomic structure calculation.

Similar ion-like $3d^x 4s^0$ configurations occur in transition metal complexes as described by the simple crystal field theory, even if the metal has oxidation state 0. For example, chromium hexacarbonyl can be described as a chromium atom (not ion) surrounded by six carbon monoxide ligands. The electron configuration of the central chromium atom is described as $3d^6$ with the six electrons filling the three lower-energy d orbitals between the ligands. The other two d orbitals are at higher energy due to the crystal field of the ligands. This picture is consistent with the experimental fact that the complex is diamagnetic, meaning that it has no unpaired electrons. However, in a more accurate description using molecular orbital theory, the d-like orbitals occupied by the six electrons are no longer identical with the d orbitals of the free atom.

Other Exceptions to Madelung's Rule

There are several more exceptions to Madelung's rule among the heavier elements, and it is more and more difficult to resort to simple explanations, such as the stability of half-filled subshells. It is possible to predict most of the exceptions by Hartree–Fock calculations, which are an approximate method for taking account of the effect of the other electrons on orbital energies. For the heavier elements, it is also necessary to take account of the effects of Special Relativity on the energies of the atomic orbitals, as the inner-shell electrons are moving at speeds approaching the speed of light. In general, these relativistic effects tend to decrease the energy of the s-orbitals in relation to the other atomic orbitals. The table below shows the ground state configuration in terms of orbital occupancy, but it does not show the ground state in terms of the sequence of orbital energies as determined spectroscopically. For example, in the transition metals, the 4s orbital is of a higher energy than the 3d orbitals; and in the lanthanides, the 6s is higher than the 4f and 5d. The ground states can be seen in the Electron configurations of the elements (data page).

Electron shells filled in violation of Madelung's rule (red)											
Period 4			Period 5			Period 6			Period 7		
Element	Z	Electron Configuration	Element	Z	Electron Configuration	Element	Z	Electron Configuration	Element	Z	Electron Configuration
						Lanthanum	57	[Xe] $6s^2$ $5d^1$	Actinium	89	[Rn] $7s^2$ $6d^1$
						Cerium	58	[Xe] $6s^2$ $4f^1$ $5d^1$	Thorium	90	[Rn] $7s^2$ $6d^2$

Element	No.	Config	Element	No.	Config	Element	No.	Config	Element	No.	Config
						Praseodymium	59	$[Xe]\,6s^2\,4f^3$	Protactinium	91	$[Rn]\,7s^2\,5f^2\,6d^1$
						Neodymium	60	$[Xe]\,6s^2\,4f^4$	Uranium	92	$[Rn]\,7s^2\,5f^3\,6d^1$
						Promethium	61	$[Xe]\,6s^2\,4f^5$	Neptunium	93	$[Rn]\,7s^2\,5f^4\,6d^1$
						Samarium	62	$[Xe]\,6s^2\,4f^6$	Plutonium	94	$[Rn]\,7s^2\,5f^6$
						Europium	63	$[Xe]\,6s^2\,4f^7$	Americium	95	$[Rn]\,7s^2\,5f^7$
						Gadolinium	64	$[Xe]\,6s^2\,4f^7\,5d^1$	Curium	96	$[Rn]\,7s^2\,5f^7\,6d^1$
						Terbium	65	$[Xe]\,6s^2\,4f^9$	Berkelium	97	$[Rn]\,7s^2\,5f^9$
Scandium	21	$[Ar]\,4s^2\,3d^1$	Yttrium	39	$[Kr]\,5s^2\,4d^1$	Lutetium	71	$[Xe]\,6s^2\,4f^{14}\,5d^1$	Lawrencium	103	$[Rn]\,7s^2\,5f^{14}\,7p^1$
Titanium	22	$[Ar]\,4s^2\,3d^2$	Zirconium	40	$[Kr]\,5s^2\,4d^2$	Hafnium	72	$[Xe]\,6s^2\,4f^{14}\,5d^2$	Rutherfordium	104	$[Rn]\,7s^2\,5f^{14}\,6d^2$
Vanadium	23	$[Ar]\,4s^2\,3d^3$	Niobium	41	$[Kr]\,5s^1\,4d^4$	Tantalum	73	$[Xe]\,6s^2\,4f^{14}\,5d^3$	Dubnium	105	$[Rn]\,7s^2\,5f^{14}\,6d^3$
Chromium	24	$[Ar]\,4s^1\,3d^5$	Molybdenum	42	$[Kr]\,5s^1\,4d^5$	Tungsten	74	$[Xe]\,6s^2\,4f^{14}\,5d^4$	Seaborgium	106	$[Rn]\,7s^2\,5f^{14}\,6d^4$
Manganese	25	$[Ar]\,4s^2\,3d^5$	Technetium	43	$[Kr]\,5s^2\,4d^5$	Rhenium	75	$[Xe]\,6s^2\,4f^{14}\,5d^5$	Bohrium	107	$[Rn]\,7s^2\,5f^{14}\,6d^5$
Iron	26	$[Ar]\,4s^2\,3d^6$	Ruthenium	44	$[Kr]\,5s^1\,4d^7$	Osmium	76	$[Xe]\,6s^2\,4f^{14}\,5d^6$	Hassium	108	$[Rn]\,7s^2\,5f^{14}\,6d^6$
Cobalt	27	$[Ar]\,4s^2\,3d^7$	Rhodium	45	$[Kr]\,5s^1\,4d^8$	Iridium	77	$[Xe]\,6s^2\,4f^{14}\,5d^7$			
Nickel	28	$[Ar]\,4s^2\,3d^8$ or $[Ar]\,4s^1\,3d^9$ (disputed)	Palladium	46	$[Kr]\,4d^{10}$	Platinum	78	$[Xe]\,6s^1\,4f^{14}\,5d^9$			

Cop-per	29	[Ar] 3d^{10} 4s^1	Silver	47	[Kr] 5s^1 4d^{10}	Gold	79	[Xe] 6s^1 4f^{14} 5d^{10}	
Zinc	30	[Ar] 3d^{10} 4s^2	Cadmium	48	[Kr] 5s^2 4d^{10}	Mercury	80	[Xe] 6s^2 4f^{14} 5d^{10}	

The electron-shell configuration of elements beyond hassium has not yet been empirically verified, but they are expected to follow Madelung's rule without exceptions until element 120.

Electron Configuration in Molecules

In molecules, the situation becomes more complex, as each molecule has a different orbital structure. The molecular orbitals are labelled according to their symmetry, rather than the atomic orbital labels used for atoms and monatomic ions: hence, the electron configuration of the dioxygen molecule, O_2, is written $1\sigma_g^2 \, 1\sigma_u^2 \, 2\sigma_g^2 \, 2\sigma_u^2 \, 3\sigma_g^2 \, 1\pi_u^4 \, 1\pi_g^2$, or equivalently $1\sigma_g^2 \, 1\sigma_u^2 \, 2\sigma_g^2 \, 2\sigma_u^2 \, 1\pi_u^4 \, 3\sigma_g^2 \, 1\pi_g^2$. The term $1\pi_g^2$ represents the two electrons in the two degenerate π^*-orbitals (antibonding). From Hund's rules, these electrons have parallel spins in the ground state, and so dioxygen has a net magnetic moment (it is paramagnetic). The explanation of the paramagnetism of dioxygen was a major success for molecular orbital theory.

The electronic configuration of polyatomic molecules can change without absorption or emission of a photon through vibronic couplings.

Electron Configuration in Solids

In a solid, the electron states become very numerous. They cease to be discrete, and effectively blend into continuous ranges of possible states (an electron band). The notion of electron configuration ceases to be relevant, and yields to band theory.

Applications

The most widespread application of electron configurations is in the rationalization of chemical properties, in both inorganic and organic chemistry. In effect, electron configurations, along with some simplified form of molecular orbital theory, have become the modern equivalent of the valence concept, describing the number and type of chemical bonds that an atom can be expected to form.

This approach is taken further in computational chemistry, which typically attempts to make quantitative estimates of chemical properties. For many years, most such calculations relied upon the "linear combination of atomic orbitals" (LCAO) approximation, using an ever larger and more complex basis set of atomic orbitals as the starting point. The last step in such a calculation is the assignment of electrons among the molecular orbitals according to the Aufbau principle. Not all methods in calculational chemistry rely on electron configuration: density functional theory (DFT) is an important example of a method that discards the model.

For atoms or molecules with more than one electron, the motion of electrons are correlated and such a picture is no longer exact. A very large number of electronic configurations are needed to exactly describe any multi-electron system, and no energy can be associated with one single configuration. However, the electronic wave function is usually dominated by a very small number of configurations and therefore the notion of electronic configuration remains essential for multi-electron systems.

A fundamental application of electron configurations is in the interpretation of atomic spectra. In this case, it is necessary to supplement the electron configuration with one or more term symbols, which describe the different energy levels available to an atom. Term symbols can be calculated for any electron configuration, not just the ground-state configuration listed in tables, although not all the energy levels are observed in practice. It is through the analysis of atomic spectra that the ground-state electron configurations of the elements were experimentally determined.

Molecular Symmetry

Molecular symmetry in chemistry describes the symmetry present in molecules and the classification of molecules according to their symmetry. Molecular symmetry is a fundamental concept in chemistry, as it can predict or explain many of a molecule's chemical properties, such as its dipole moment and its allowed spectroscopic transitions (based on selection rules such as the Laporte rule). Many university level textbooks on physical chemistry, quantum chemistry, and inorganic chemistry devote a chapter to symmetry.

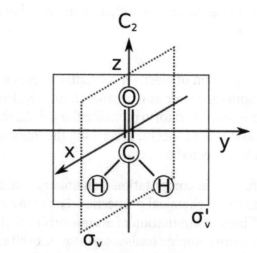

Symmetry elements of formaldehyde. C_2 is a two-fold rotation axis. σ_v and σ_v' are two non-equivalent reflection planes.

While various frameworks for the study of molecular symmetry exist, group theory is the predominant one. This framework is also useful in studying the symmetry of molecular orbitals, with applications such as the Hückel method, ligand field theory, and the Woodward-Hoffmann rules. Another framework on a larger scale is the use of crystal systems to describe crystallographic symmetry in bulk materials.

Many techniques for the practical assessment of molecular symmetry exist, including X-ray crystallography and various forms of spectroscopy, for example infrared spectroscopy of metal carbonyls. Spectroscopic notation is based on symmetry considerations.

Symmetry Concepts

The study of symmetry in molecules is an adaptation of mathematical group theory.

	Examples of the relationship between chirality and symmetry		
Rotational axis (C_n)	**Improper rotational elements (S_n)**		
	Chiral no S_n	**Achiral** mirror plane $S_1 = \sigma$	**Achiral** inversion centre $S_2 = i$
C_1			
C_2			

Elements

The symmetry of a molecule can be described by 5 types of symmetry elements.

- Symmetry axis: an axis around which a rotation by $\frac{360°}{n}$ results in a molecule indistinguishable from the original. This is also called an n-fold rotational axis and abbreviated C_n. Examples are the C_2 in water and the C_3 in ammonia. A molecule can have more than one symmetry axis; the one with the highest n is called the principal axis, and by convention is assigned the z-axis in a Cartesian coordinate system.

- Plane of symmetry: a plane of reflection through which an identical copy of the original molecule is given. This is also called a mirror plane and abbreviated σ. Water has two of them: one in the plane of the molecule itself and one perpendicular to it. A symmetry plane parallel with the principal axis is dubbed *vertical* (σ_v) and one perpendicular to it *horizontal* (σ_h). A third type of symmetry plane exists: If a vertical symmetry plane additionally bisects the angle between two 2-fold rotation axes perpendicular to the principal axis, the plane is dubbed dihedral (σ_d). A symmetry plane can also be identified by its Cartesian orientation, e.g., (xz) or (yz).

- Center of symmetry or inversion center, abbreviated *i*. A molecule has a center of symmetry when, for any atom in the molecule, an identical atom exists diametrically opposite this center an equal distance from it. In other words, a molecule has a center of symmetry when the points (x,y,z) and (-x,-y,-z) correspond to identical objects. For example, if there is an oxygen atom in some point (x,y,z), then there is an oxygen atom in the point (-x,-y,-z). There may or may not be an atom at the center. Examples are xenon tetrafluoride where the inversion center is at the Xe atom, and benzene (C_6H_6) where the inversion center is at the center of the ring.

- Rotation-reflection axis: an axis around which a rotation by $\frac{360°}{n}$, followed by a reflection in a plane perpendicular to it, leaves the molecule unchanged. Also called an n-fold improper rotation axis, it is abbreviated S_n. Examples are present in tetrahedral silicon tetrafluoride, with three S_4 axes, and the staggered conformation of ethane with one S_6 axis.

- Identity, abbreviated to E, from the German 'Einheit' meaning unity. This symmetry element simply consists of no change: every molecule has this element. While this element seems physically trivial, it must be included in the list of symmetry elements so that they form a mathematical group, whose definition requires inclusion of the identity element. It is so called because it is analogous to multiplying by one (unity). In other words, E is a property that any object needs to have regardless of its symmetry properties.

Operations

XeF$_4$, with square planar geometry, has 1 C$_4$ axis and 4 C$_2$ axes orthogonal to C$_4$. These five axes plus the mirror plane perpendicular to the C$_4$ axis define the D$_{4h}$ symmetry group of the molecule.

The 5 symmetry elements have associated with them 5 types of symmetry operations, which leave the molecule in a state indistinguishable from the starting state. They are sometimes distinguished from symmetry elements by a caret or circumflex. Thus, $\hat{C}_n$ is the rotation of a molecule around an axis and $\hat{E}$ is the identity operation. A symmetry element can have more than one symmetry operation associated with it. For example, the C_4 axis of the square xenon tetrafluoride (XeF_4) molecule is associated with two $\hat{C}_4$ rotations (90°) in opposite directions and a $\hat{C}_2$ rotation (180°). Since $\hat{C}_1$ is equivalent to $\hat{E}$, $\hat{S}_1$ to σ and $\hat{S}_2$ to $\hat{\imath}$, all symmetry operations can be classified as either proper or improper rotations.

Molecular Symmetry Groups

Groups

The symmetry operations of a molecule (or other object) form a *group*, which is a mathematical structure usually denoted in the form $(G,*)$ consisting of a set G and a binary combination operation say '*' satisfying certain properties listed below.

In a molecular symmetry group, the group elements are the symmetry operations (*not* the symmetry elements), and the binary combination consists of applying first one symmetry operation and then the other. An example is the sequence of a C_4 rotation about the z-axis and a reflection in the xy-plane, denoted $\sigma(xy)C_4$. By convention the order of operations is from right to left.

A molecular symmetry group obeys the defining properties of any group.

(1) *closure* property:

For every pair of elements x and y in G, the *product* $x*y$ is also in G.

(in symbols, for every two elements $x, y \in G$, $x*y$ is also in G).

This means that the group is *closed* so that combining two elements produces no new elements. Symmetry operations have this property because a sequence of two operations will produce a third state indistinguishable from the second and therefore from the first, so that the net effect on the molecule is still a symmetry operation.

(2) *associative* property:

For every x and y and z in G, both $(x*y)*z$ and $x*(y*z)$ result with the same element in G.

(in symbols, $(x*y)*z = x*(y*z)$ for every x, y, and $z \in G$)

(3) *existence of identity* property:

There must be an element (say e) in G such that product any element of G with e make no change to the element.

(in symbols, $x*e=e*x= x$ for every $x \in G$)

(4) *existence of inverse* property:

For each element (x) in G, there must be an element y in G such that product of x and y is the identity element e.

(in symbols, for each $x \in G$ there is a $y \in G$ such that $x*y=y*x= e$ for every $x \in G$)

The *order* of a group is the number of elements in the group. For groups of small orders, the group properties can be easily verified by considering its composition table, a table whose rows and columns correspond to elements of the group and whose entries correspond to their products.

Point Group

The successive application (or *composition*) of one or more symmetry operations of a molecule has an effect equivalent to that of some single symmetry operation of the molecule. For example, a C_2 rotation followed by a σ_v reflection is seen to be a σ_v' symmetry operation: $\sigma_v * C_2 = \sigma_v'$. (Note that "Operation A followed by B to form C" is written BA = C). Moreover, the set of all symmetry operations (including this composition operation) obeys all the properties of a group, given above. So $(S, *)$ is a group, where S is the set of all symmetry operations of some molecule, and * denotes the composition (repeated application) of symmetry operations.

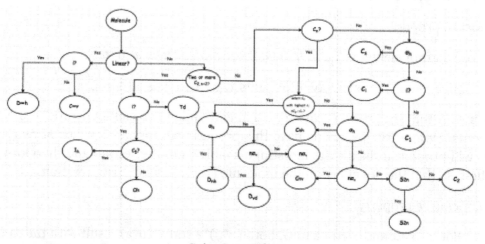

Point Group Chart

This group is called the point group of that molecule, because the set of symmetry operations leave at least one point fixed (though for some symmetries an entire axis or an entire plane remains fixed). In other words, a point group is a group that summarizes all symmetry operations that all molecules in that category have. The symmetry of a crystal, by contrast, is described by a space group of symmetry operations, which includes translations in space.

Examples

Assigning each molecule a point group classifies molecules into categories with similar symmetry properties. For example, PCl_3, POF_3, XeO_3, and NH_3 all share identical symmetry operations. They all can undergo the identity operation E, two different C_3 rotation operations, and three different σ_v plane reflections without altering their identities, so they are placed in one point group, C_{3v}, with order 6. Similarly, water (H_2O) and hydrogen sulfide (H_2S) also share identical symmetry operations. They both undergo the identity operation E, one C_2 rotation, and two σ_v reflections without altering their identities, so they are both placed in one point group, C_{2v}, with order 4. This classification system helps scientists to study molecules more efficiently, since molecules in the same point group tend to exhibit similar bonding schemes, molecular bonding diagrams, and spectroscopic properties.

Common Point Groups

The following table contains a list of point groups labelled using the Schoenflies notation which is common in chemistry and molecular spectroscopy. The description of structure includes common shapes of molecules based on VSEPR theory.

Point group	Symmetry operations	Simple description of typical geometry	Example 1	Example 2	Example 3
C_1	E	no symmetry, chiral	bromochlorofluoromethane	lysergic acid	
C_s	E σ_h	mirror plane, no other symmetry	thionyl chloride	hypochlorous acid	chloroiodomethane
C_i	E i	inversion center	(S,R) 1,2-dibromo-1,2-dichloroethane (*anti* conformer)		

$C_{\infty v}$	E $2C_\infty$ $\infty\sigma_v$	linear	Hydrogen fluoride	nitrous oxide (dinitrogen monoxide)	
$D_{\infty h}$	E $2C_\infty$ $\infty\sigma_i$ i $2S_\infty$ ∞C_2	linear with inversion center	oxygen	carbon dioxide	
C_2	E C_2	"open book geometry," chiral	peroxide		
C_3	E C_3	propeller, chiral	triphenylphosphine		
C_{2h}	E C_2 i σ_h	planar with inversion center	trans-1,2-dichloroethylene		
C_{3h}	E C_3 C_3^2 σ_h S_3 S_3^5	propeller	boric acid		

C_{2v}	E $\quad$ C$_2$ $\sigma_v(xz)$ $\sigma_v'(yz)$	angular (H$_2$O) or see-saw (SF$_4$)	water	sulfur tetrafluoride	sulfuryl fluoride
C_{3v}	E $\quad$ 2C$_3$ 3σ_v	trigonal py-ramidal	ammonia	phosphorus oxychloride	
C_{4v}	E $\quad$ 2C$_4$ C$_2$ $\quad$ 2σ_v 2σ_d	square py-ramidal	xenon oxytetrafluoride		
D_2	E C$_2$(x) C$_2$(y) C$_2$(z)	twist, chiral	cyclohexane twist conformation		
D_3	E C$_3$(z) 3C$_2$	triple helix, chiral	chloride anions omitted Tris(ethylenediamine) cobalt(III) cation		
D_{2h}	E C$_2$(z) C$_2$(y) C$_2$(x) i σ(xy) σ(xz) σ(yz)	planar with inversion center	ethylene	dinitrogen tetroxide	diborane

D_{3h}	E $2C_3$ $3C_2$ σ_h $2S_3$ $3\sigma_v$	trigonal planar or trigonal bi-pyramidal	boron trifluoride	phosphorus pentachloride	
D_{4h}	E $2C_4$ C_2 $2C_2'$ $2C_2''$ i $2S_4$ σ_h $2\sigma_v$ $2\sigma_d$	square planar	xenon tetrafluoride	octachloro dimolybdate(II) anion	
D_{5h}	E $2C_5$ $2C_5^2$ $5C_2$ σ_h $2S_5$ $2S_5^3$ $5\sigma_v$	pentagonal	ruthenocene	C_{70}	
D_{6h}	E $2C_6$ $2C_3$ C_2 $3C_2'$ $3C_2''$ i $2S_3$ $2S_6$ σ_h $3\sigma_d$ $3\sigma_v$	hexagonal	benzene	bis(benzene) chromium	
D_{2d}	E $2S_4$ C_2 $2C_2'$ $2\sigma_d$	90° twist	allene	tetrasulfur tetranitride	

D_{3d}	E 2C_3 3C_2 i 2S_6 3σ_d	60° twist	ethane (staggered rotamer)	cyclohexane chair conformation	
D_{4d}	E 2S_8 2C_4 2S_8^3 C_2 4C_2' 4σ_d	45° twist	dimanganese decacarbonyl (staggered rotamer)		
D_{5d}	E 2C_5 2C_5^2 5C_2 i 3S_{10}^3 2S_{10} 5σ_d	36° twist	ferrocene (staggered rotamer)		
T_d	E 8C_3 3C_2 6S_4 6σ_d	tetrahedral	methane	phosphorus pentoxide	adamantane
O_h	E 8C_3 6C_2 6C_4 3C_2 i 6S_4 8S_6 3σ_h 6σ_d	octahedral or cubic	cubane	sulfur hexafluoride	
I_h	E 12C_5 12C_5^2 20C_3 15C_2 i 12S_{10} 12S_{10}^3 20S_6 15σ	icosahedral or dodecahedral	Buckminsterfullerene	dodecaborate anion	dodecahedrane

Representations

The symmetry operations can be represented in many ways. A convenient representation is by matrices. For any vector representing a point in Cartesian coordinates, left-multiplying it gives the new location of the point transformed by the symmetry operation. Composition of operations corresponds to matrix multiplication. Within a point group, a multiplication of the matrices of two symmetry operations leads to a matrix of another symmetry operation in the same point group. For example, in the C_{2v} example this is:

$$\underbrace{\begin{bmatrix} -1 & 0 & 0 \\ 0 & -1 & 0 \\ 0 & 0 & 1 \end{bmatrix}}_{C_2} \times \underbrace{\begin{bmatrix} 1 & 0 & 0 \\ 0 & -1 & 0 \\ 0 & 0 & 1 \end{bmatrix}}_{\sigma_v} = \underbrace{\begin{bmatrix} -1 & 0 & 0 \\ 0 & 1 & 0 \\ 0 & 0 & 1 \end{bmatrix}}_{\sigma'_v}$$

Although an infinite number of such representations exist, the irreducible representations (or "irreps") of the group are commonly used, as all other representations of the group can be described as a linear combination of the irreducible representations.

Character Tables

For each point group, a character table summarizes information on its symmetry operations and on its irreducible representations. As there are always equal numbers of irreducible representations and classes of symmetry operations, the tables are square.

The table itself consists of characters that represent how a particular irreducible representation transforms when a particular symmetry operation is applied. Any symmetry operation in a molecule's point group acting on the molecule itself will leave it unchanged. But, for acting on a general entity, such as a vector or an orbital, this need not be the case. The vector could change sign or direction, and the orbital could change type. For simple point groups, the values are either 1 or −1: 1 means that the sign or phase (of the vector or orbital) is unchanged by the symmetry operation (*symmetric*) and −1 denotes a sign change (*asymmetric*).

The representations are labeled according to a set of conventions:

- A, when rotation around the principal axis is symmetrical
- B, when rotation around the principal axis is asymmetrical
- E and T are doubly and triply degenerate representations, respectively
- when the point group has an inversion center, the subscript g (German: *gerade*

or even) signals no change in sign, and the subscript u (*ungerade* or uneven) a change in sign, with respect to inversion.

- with point groups $C_{\infty v}$ and $D_{\infty h}$ the symbols are borrowed from angular momentum description: Σ, Π, Δ.

The tables also capture information about how the Cartesian basis vectors, rotations about them, and quadratic functions of them transform by the symmetry operations of the group, by noting which irreducible representation transforms in the same way. These indications are conventionally on the righthand side of the tables. This information is useful because chemically important orbitals (in particular p and d orbitals) have the same symmetries as these entities.

The character table for the C_{2v} symmetry point group is given below:

C_{2v}	E	C_2	$\sigma_v(xz)$	$\sigma_v'(yz)$		
A_1	1	1	1	1	z	x^2, y^2, z^2
A_2	1	1	−1	−1	R_z	xy
B_1	1	−1	1	−1	x, R_y	xz
B_2	1	−1	−1	1	y, R_x	yz

Consider the example of water (H_2O), which has the C_{2v} symmetry described above. The $2p_x$ orbital of oxygen has B_1 symmetry as in the fourth row of the character table above, with x in the sixth column). It is oriented perpendicular to the plane of the molecule and switches sign with a C_2 and a $\sigma_v'(yz)$ operation, but remains unchanged with the other two operations (obviously, the character for the identity operation is always +1). This orbital's character set is thus {1, −1, 1, −1}, corresponding to the B_1 irreducible representation. Likewise, the $2p_z$ orbital is seen to have the symmetry of the A_1 irreducible representation, $2p_y$ B_2, and the $3d_{xy}$ orbital A_2. These assignments and others are noted in the rightmost two columns of the table.

Historical Background

Hans Bethe used characters of point group operations in his study of ligand field theory in 1929, and Eugene Wigner used group theory to explain the selection rules of atomic spectroscopy. The first character tables were compiled by László Tisza (1933), in connection to vibrational spectra. Robert Mulliken was the first to publish character tables in English (1933), and E. Bright Wilson used them in 1934 to predict the symmetry of vibrational normal modes. The complete set of 32 crystallographic point groups was published in 1936 by Rosenthal and Murphy.

Non-rigid Molecules

The symmetry groups described above are useful for describing *rigid* molecules which undergo only small oscillations about a single equilibrium geometry, so that the symmetry operations all correspond to simple geometrical operations. However Longuet-Higgins has proposed a more general type of symmetry groups suitable for non-rigid molecules with multiple equivalent geometries. These groups are known as *permutation-inversion* groups, because a symmetry operation may be an energetically feasible permutation of equivalent nuclei, or an inversion with respect to the centre of mass, or a combination of the two.

For example, ethane (C_2H_6) has three equivalent staggered conformations. Conversion of one conformation to another occurs at ordinary temperature by *internal rotation* of one methyl group relative to the other. This is not a rotation of the entire molecule about the C_3 axis, but can be described as a permutation of the three identical hydrogens of one methyl group. Although each conformation has D_{3d} symmetry as in the table above, description of the internal rotation and associated quantum states and energy levels requires the more complete permutation-inversion group.

Similarly, ammonia (NH_3) has two equivalent pyramidal (C_{3v}) conformations which are interconverted by the process known as nitrogen inversion. This is not an inversion in the sense used for symmetry operations of rigid molecules, since NH_3 has no inversion center. Rather it is a reflection of all atoms about the centre of mass (close to the nitrogen), which happens to be energetically feasible for this molecule. Again the permutation-inversion group is used to describe the interaction of the two geometries.

A second and similar approach to the symmetry of nonrigid molecules is due to Altmann. In this approach the symmetry groups are known as *Schrödinger supergroups* and consist of two types of operations (and their combinations): (1) the geometric symmetry operations (rotations, reflections, inversions) of rigid molecules, and (2) *isodynamic operations* which take a nonrigid molecule into an energetically equivalent form by a physically reasonable process such as rotation about a single bond (as in ethane) or a molecular inversion (as in ammonia).

Interatomic Potential

Interatomic potentials are mathematical functions for calculating the potential energy of a system of atoms with given positions in space. Interatomic potentials are widely used as the physical basis of molecular mechanics and molecular dynamics simulations in chemistry, molecular physics and materials physics, sometimes in connection with such effects as cohesion, thermal expansion and elastic properties of materials.

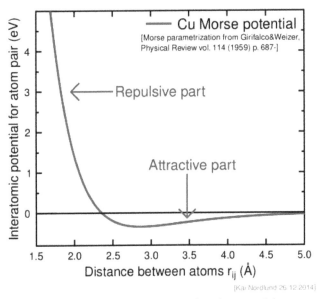

Typical shape of an interatomic pair potential.

Functional form

Interatomic potentials can be written as a series expansion of functional terms that depend on the position of one, two, three, etc. atoms at a time. Then the total energy of the system V can be written as

$$V_{TOT} = \sum_i^N V_1(\vec{r}_i) + \sum_{i,j}^N V_2(\vec{r}_i,\vec{r}_j) + \sum_{i,j,k}^N V_3(\vec{r}_i,\vec{r}_j,\vec{r}_k) + \cdots$$

Here V_1 is the one-body term, V_2 the two-body term, V_3 the three body term, N the number of atoms in the system, $\vec{r}_i$ the position of atom i, etc. i, j and k are indices that loop over atom positions.

Note that in case the pair potential is given per atom pair, in the two-body term the potential should be multiplied by 1/2 as otherwise each bond is counted twice, and similarly the three-body term by 1/6. Alternatively, the summation of the pair term can be restricted to cases $i < j$ and similarly for the three-body term $i < j < k$, if the potential form is such that it is symmetric with respect to exchange of the j and k indices (this may not be the case for potentials for multielemental systems).

The one-body term is only meaningful if the atoms are in an external field (e.g. an electric field). In the absence of external fields, the potential V should not depend on the absolute position of atoms, but only on the relative positions. This means that the functional form can be rewritten as a function of interatomic distances $r_{ij} = |\vec{r}_i - \vec{r}_j|$ and angles between the bonds (vectors to neighbours) θ_{ijk}. Then, in the absence of external forces, the general form becomes

$$V_{TOT} = \sum_{i,j}^{N} V_2(r_{ij}) + \sum_{i,j,k}^{N} V_3(r_{ij}, r_{ik}, \theta_{ijk}) + \cdots$$

In the three-body term V_3 the interatomic distance r_{jk} is not needed since the three terms $r_{ij}, r_{ik}, \theta_{ijk}$ are sufficient to give the relative positions of three atoms i,j,k in three-dimensional space. Any terms of order higher than 2 are also called many-body potentials. In some interatomic potentials the manybody interactions are embedded into the terms of a pair potential

In principle the sums in the expressions run over all N atoms. However, if the range of the interatomic potential is finite, i.e. the potentials $V(r) \equiv 0$ above some cutoff distance r_{cut}, the summing can be restricted to atoms within the cutoff distance of each other. By also using a cellular method for finding the neighbours, the MD algorithm can be an O(N) algorithm. Potentials with an infinite range can be summed up efficiently by Ewald summation and its further developments.

Force Calculation

The forces acting between atoms can be obtained by differentiation of the total energy with respect to atom positions. That is, to get the force on atom i one should take the three-dimensional derivative (gradient) with respect to the position of atom i:

$$\vec{F}_i = \nabla_{\vec{r}_i} V_{TOT}$$

For two-body potentials this gradient reduces, thanks to the symmetry with respect to ij in the potential form, to straightforward differentiation with respect to the interatomic distances r_{ij}. However, for many-body potentials (three-body, four-body, etc.) the differentiation becomes considerably more complex   since the potential may not be any longer symmetric with respect to ij exchange. In other words, also the energy of atoms k that are not direct neighbours of i can depend on the position $\vec{r}_i$ because of angular and other many-body terms, and hence contribute to the gradient $\nabla_{\vec{r}_i}$.

Classes of Interatomic Potentials

Interatomic potentials come in many different varieties, with different physical motivations. Even for single well-known elements such as silicon, a wide variety of potentials quite different in functional form and motivation have been developed. The true interatomic interactions are quantum mechanical in nature, and there is no known way in which the true interactions described by the Schrödinger equation or Dirac equation for all electrons and nuclei could be cast into an analytical functional form. Hence all analytical interatomic potentials are by necessity approximations.

Pair Potentials

The arguably simplest widely used interatomic interaction model is the Lennard-Jones potential

$$V_{LJ} = 4\varepsilon \left[\left(\frac{\sigma}{r} \right)^{12} - \left(\frac{\sigma}{r} \right)^{6} \right]$$

where ε is the depth of the potential well and σ is the distance at which the potential crosses zero. The term proportional to $1/r^6$ in the potential can be motivated from a classical or quantum mechanical description of the interaction between induced electric dipoles. This potential seems to be quite accurate for noble gases, and is widely used for systems where dipole interactions are significant, including in chemistry force fields to describe intermermolecular interactions.

Another simple and widely used pair potential is the Morse potential, which consists simply of a sum of two exponentials.

$$V(r) = D_e (e^{-2a(r-r_e)} - 2e^{-a(r-r_e)})$$

Here D_e is the equilibrium bond energy and r_e the bond distance. The Morse potential has been applied to studies of molecular vibrations and solids , and although rarely used anymore, inspired the functional form of more modern potentials such as the bond-order potentials.

Ionic materials are often described by a sum of a short-range repulsive term, such as the Buckingham pair potential, and a long-range Coulomb potential giving the ionic interactions between the ions forming the material. The short-range term for ionic materials can also be of many-body character .

Pair potentials have some inherent limitations, like the inability to describe all 3 elastic constants of cubic metals. Hence modern molecular dynamics simulations are to a large extent carried out with different kinds of many-body potentials.

Many-body Potentials

The Stilinger-Weber potential is a potential that has a two-body and three-body terms of the standard form

$$V_{TOT} = \sum_{i,j}^{N} V_2(r_{ij}) + \sum_{i,j,k}^{N} V_3(r_{ij}, r_{ik}, \theta_{ijk})$$

where the three-body term describes how the potential energy changes with bond bending. It was originally developed for pure Si, but has been extended to many other elements and compounds   and also formed the basis for other Si potentials.

Metals are very commonly described with what can be called "EAM-like" potentials, i.e. potentials that share the same functional form as the embedded atom model. In these potentials, the total potential energy is written

$$V_{TOT} = \sum_{i}^{N} F_i \left(\sum_{j} \rho(r_{ij}) \right) + \frac{1}{2} \sum_{i,j}^{N} V_2(r_{ij})$$

where F_i is a so-called embedding function (not to be confused with the force $\vec{F}_i$) that is a function of the sum of the so-called electron density $\rho(r_{ij})$. V_2 is a pair potential that usually is purely repulsive. In the original formulation the electron density function $\rho(r_{ij})$ was obtained from true atomic electron densities, and the embedding function was motivated from density-functional theory as the energy needed to 'embed' an atom into the electron density. . However, many other potentials used for metals share the same functional form but motivate the terms differently, e.g. based on tight-binding theory or other motivations.

EAM-like potentials are usually implemented as numerical tables. A collection of tables is available at the interatomic potential repository at NIST

Covalently bonded materials are often described by bond order potentials, sometimes also called Tersoff-like or Brenner-like potentials. These have in general a form that resembles a pair potential:

$$V_{ij}(r_{ij}) = V_{repulsive}(r_{ij}) + b_{ijk} V_{attractive}(r_{ij})$$

where the repulsive and attractive part are simple exponential functions similar to those in the Morse potential. However, the strength is modified by the environment of the atom i via the b_{ijk} term. If implemented without an explicit angular dependence, these potentials can be shown to be mathematically equivalent to some varieties of EAM-like potentials Thanks to this equivalence, the bond-order potential formalism has been implemented also for many metal-covalent mixed materials.

Repulsive Potentials for Short-range Interactions

For very short interatomic separations, important in radiation material science, the interactions can be described quite accurately with screened Coulomb potentials which have the general form

$$V(r_{ij}) = \frac{1}{4\pi\varepsilon_0} \frac{Z_1 Z_2 e^2}{r_{ij}} \varphi(r/a)$$

here $\varphi(r) \to 1$ when $r \to 0$. Here Z_1 and Z_2 are the charges of the interacting nuclei, and a is the so-called screening parameter. A widely used popular screening function is the

"Universal ZBL" one. and more accurate ones can be obtained from all-electron quantum chemistry calculations In binary collision approximation simulations this kind of potential can be used to describe the nuclear stopping power.

Potential Fitting

Since the interatomic potentials are approximations, they by necessity all involve parameters that need to be adjusted to some reference values. In simple potentials such as the Lennard-Jones and Morse ones, the parameters can be set directly to match e.g. the equilibrium bond length and bond strength of a dimer molecule or the cohesive energy of a solid . However, many-body potentials often contain tens or even hundreds of adjustable parameters. These can be fit into a larger set of experimental data, or materials properties derived from more fundamental simulation models such as density-functional theory. For solids, a well-constructed many-body potential can often describe at least the equilibrium crystal structure cohesion and lattice constant, linear elastic constants, and basic point defect properties of all the elements and stable compounds well. The aim of most potential construction and fitting is to make the potential transferable, i.e. that it can describe materials properties that are clearly dif-ferent from those it was fitted. As an example of demonstrated partial transferability, a review of interatomic potentials of Si found that for instance the Stillinger-Weber and Tersoff III potentials for Si are indeed able to describe several (but certainly not all) materials properties they were not fitted to .

The NIST interatomic potential repository provides a collection of fitted interatomic potentials, either as fitted parameter values or numerical tables of the potential functions.

Reliability of Interatomic Potentials

Classical interatomic potentials cannot reproduce all phenomena. Sometimes quantum description is necessary. Density functional theory is used to overcome this limitation.

Quantum Harmonic Oscillator

The quantum harmonic oscillator is the quantum-mechanical analog of the classical harmonic oscillator. Because an arbitrary potential can usually be approximated as a harmonic potential at the vicinity of a stable equilibrium point, it is one of the most important model systems in quantum mechanics. Furthermore, it is one of the few quantum-mechanical systems for which an exact, analytical solution is known.

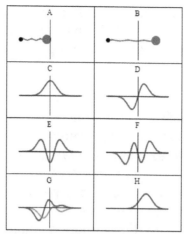

Some trajectories of a harmonic oscillator according to Newton's laws of classical mechanics (A–B), and according to the Schrödinger equation of quantum mechanics (C–H). In A–B, the particle (represented as a ball attached to a spring) oscillates back and forth. In C–H, some solutions to the Schrödinger Equation are shown, where the horizontal axis is position, and the vertical axis is the real part (blue) or imaginary part (red) of the wavefunction. C, D, E, F, but not G, H, are energy eigenstates. H is a coherent state—a quantum state that approximates the classical trajectory.

One-dimensional Harmonic Oscillator

Hamiltonian and Energy Eigenstates

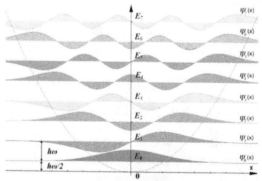

Wavefunction representations for the first eight bound eigenstates, $n = 0$ to 7. The horizontal axis shows the position x. Note: The graphs are not normalized, and the signs of some of the functions differ from those given in the text.

The Hamiltonian of the particle is:

$$\hat{H} = \frac{\hat{p}^2}{2m} + \frac{1}{2}m\omega^2\hat{x}^2,$$

where m is the particle's mass, ω is the angular frequency of the oscillator, $\hat{x}$ is the position operator (given by x), and $\hat{p}$ is the momentum operator, given by $\hat{p} = -i\hbar\frac{\partial}{\partial x}$. The first term in the Hamiltonian represents the possible kinetic energy states of the particle, and the second term represents its corresponding possible potential energy states.

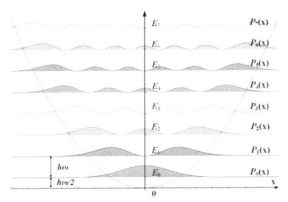

Corresponding probability densities.

One may write the time-independent Schrödinger equation,

$$\hat{H}\left|\psi\right\rangle = E\left|\psi\right\rangle,$$

where E denotes a yet-to-be-determined real number that will specify a time-independent energy level, or eigenvalue, and the solution $\left|\psi\right\rangle$ denotes that level's energy eigenstate.

One may solve the differential equation representing this eigenvalue problem in the coordinate basis, for the wave function $\langle x|\psi\rangle = \psi(x)$, using a spectral method. It turns out that there is a family of solutions. In this basis, they amount to

$$\psi_n(x) = \frac{1}{\sqrt{2^n n!}} \left(\frac{m\omega}{\pi\hbar}\right)^{1/4} \cdot e^{-\frac{m\omega x^2}{2\hbar}} \cdot H_n\left(\sqrt{\frac{m\omega}{\hbar}} x\right), \qquad n = 0,1,2,\ldots$$

The functions H_n are the physicists' Hermite polynomials,

$$H_n(z) = (-1)^n e^{z^2} \frac{d^n}{dz^n}\left(e^{-z^2}\right).$$

The corresponding energy levels are

$$E_n = \hbar\omega\left(n + \frac{1}{2}\right) = (2n+1)\frac{\hbar}{2}\omega.$$

This energy spectrum is noteworthy for three reasons. First, the energies are quantized, meaning that only discrete energy values (integer-plus-half multiples of $\hbar\omega$) are possible; this is a general feature of quantum-mechanical systems when a particle is confined. Second, these discrete energy levels are equally spaced, unlike in the Bohr model of the atom, or the particle in a box. Third, the lowest achievable energy (the energy of the $n = 0$ state, called the ground state) is not equal to the minimum of the potential well, but $\hbar\omega/2$ above it; this is called zero-point energy. Because of the zero-point energy, the position and momentum of the oscillator in the ground state are not fixed (as they would be in a classical oscillator), but have a small range of variance, in accordance with the Heisenberg uncertainty principle. This zero-point energy further has important implications in quantum field theory and quantum gravity.

The ground state probability density is concentrated at the origin, which means the particle spends most of its time at the bottom of the potential well, as one would expect for a state with little energy. As the energy increases, the probability density becomes concentrated at the classical "turning points", where the state's energy coincides with the potential energy. This is consistent with the classical harmonic oscillator, in which the particle spends most of its time (and is therefore most likely to be found) at the turning points, where it is the slowest. The correspondence principle is thus satisfied. Moreover, special nondispersive wave packets, with minimum uncertainty, called coherent states oscillate very much like classical objects, as illustrated in the figure; they are *not* eigenstates of the Hamiltonian.

Ladder Operator Method

The spectral method solution, though straightforward, is rather tedious. The "ladder operator" method, developed by Paul Dirac, allows us to extract the energy eigenvalues without directly solving the differential equation. Furthermore, it is readily generalizable to more complicated problems, notably in quantum field theory. Following this approach, we define the operators a and its adjoint $a^\dagger$,

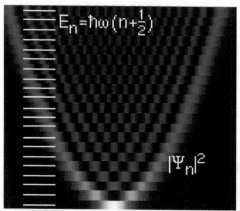

Probability densities $|\psi_n(x)|^2$ for the bound eigenstates, beginning with the ground state ($n = 0$) at the bottom and increasing in energy toward the top. The horizontal axis shows the position x, and brighter colors represent higher probability densities.

$$a = \sqrt{\frac{m\omega}{2\hbar}}\left(\hat{x} + \frac{i}{m\omega}\hat{p}\right)$$

$$a^\dagger = \sqrt{\frac{m\omega}{2\hbar}}\left(\hat{x} - \frac{i}{m\omega}\hat{p}\right)$$

This leads to the useful representation of 'x and $\hat{p}$,

$$\hat{x} = \sqrt{\frac{\hbar}{2}\frac{1}{m\omega}}(a^\dagger + a)$$

$$\hat{p} = i\sqrt{\frac{\hbar}{2}m\omega}(a^\dagger - a) .$$

The operator a is not Hermitian, since itself and its adjoint $a^\dagger$ are not equal. Yet the energy eigenstates $|n\rangle$, when operated on by these ladder operators, give

$$a^\dagger |n\rangle = \sqrt{n+1}|n+1\rangle$$

$$a|n\rangle = \sqrt{n}|n-1\rangle.$$

It is then evident that $a^\dagger$, in essence, appends a single quantum of energy to the oscillator, while a removes a quantum. For this reason, they are sometimes referred to as "creation" and "annihilation" operators.

From the relations above, we can also define a number operator N, which has the following property:

$$N = a^\dagger a$$

$$N|n\rangle = n|n\rangle.$$

The following commutators can be easily obtained by substituting the canonical commutation relation,

$$[a, a^\dagger] = 1, \qquad [N, a^\dagger] = a^\dagger, \qquad [N, a] = -a,$$

And the Hamilton operator can be expressed as

$$\hat{H} = \left(N + \frac{1}{2}\right)\hbar\omega,$$

so the eigenstate of N is also the eigenstate of energy.

The commutation property yields

$$Na^\dagger |n\rangle = \left(a^\dagger N + [N, a^\dagger]\right)|n\rangle$$

$$= \left(a^\dagger N + a^\dagger\right)|n\rangle$$

$$= (n+1)a^\dagger |n\rangle,$$

and similarly,

$$Na \mid n\rangle = (n-1)a \mid n\rangle.$$

This means that a acts on $|n\rangle$ to produce, up to a multiplicative constant, $|n-1\rangle$, and $a^{\dagger}$ acts on $|n\rangle$ to produce $|n+1\rangle$. For this reason, a is called a "lowering operator", and $a^{\dagger}$ a "raising operator". The two operators together are called ladder operators. In quantum field theory, a and $a^{\dagger}$ are alternatively called "annihilation" and "creation" operators because they destroy and create particles, which correspond to our quanta of energy.

Given any energy eigenstate, we can act on it with the lowering operator, a, to produce another eigenstate with $\hbar\omega$ less energy. By repeated application of the lowering operator, it seems that we can produce energy eigenstates down to $E = -\infty$. However, since

$$n = \langle n \mid N \mid n\rangle = \langle n \mid a^{\dagger}a \mid n\rangle = (a \mid n\rangle)^{\dagger} a \mid n\rangle \geqslant 0,$$

the smallest eigen-number is 0, and

$$a \mid 0\rangle = 0.$$

In this case, subsequent applications of the lowering operator will just produce zero kets, instead of additional energy eigenstates. Furthermore, we have shown above that

$$\hat{H} \mid 0\rangle = \frac{\hbar\omega}{2} \mid 0\rangle$$

Finally, by acting on $|0\rangle$ with the raising operator and multiplying by suitable normalization factors, we can produce an infinite set of energy eigenstates

$$\left\{ \mid 0\rangle, \mid 1\rangle, \mid 2\rangle, \ldots, \mid n\rangle, \ldots \right\},$$

such that

$$\hat{H} \mid n\rangle = \hbar\omega \left(n + \frac{1}{2} \right) \mid n\rangle,$$

which matches the energy spectrum given in the preceding section.

Arbitrary eigenstates can be expressed in terms of $|0\rangle$,

$$\mid n\rangle = \frac{(a^{\dagger})^n}{\sqrt{n!}} \mid 0\rangle.$$

Proof:

$$\langle n \mid aa^\dagger \mid n \rangle = \langle n \mid \left([a,a^\dagger] + a^\dagger a\right) \mid n \rangle = \langle n \mid (N+1) \mid n \rangle = n+1$$

$$\Rightarrow a^\dagger \mid n \rangle = \sqrt{n+1} \mid n+1 \rangle$$

$$\Rightarrow \mid n \rangle = \frac{a^\dagger}{\sqrt{n}} \mid n-1 \rangle = \frac{(a^\dagger)^2}{\sqrt{n(n-1)}} \mid n-2 \rangle = \cdots = \frac{(a^\dagger)^n}{\sqrt{n!}} \mid 0 \rangle.$$

The ground state $|0\rangle$ in the position representation is determined by $a \mid 0 \rangle = 0$,

$$\langle x \mid a \mid 0 \rangle = 0$$

$$\Rightarrow \left(x + \frac{\hbar}{m\omega} \frac{d}{dx} \right) \langle x \mid 0 \rangle = 0$$

and hence

$$\Rightarrow \langle x \mid 0 \rangle = \left(\frac{m\omega}{\pi\hbar} \right)^{\frac{1}{4}} \exp\left(-\frac{m\omega}{2\hbar} x^2 \right) = \psi_0 ,$$

$$\langle x \mid a^\dagger \mid 0 \rangle = \psi_1(x),$$

so $\psi_1(x,t) = \langle x \mid e^{-3i\omega t/2} a^\dagger \mid 0 \rangle$, and so on, as in the previous section.

Natural Length and Energy Scales

The quantum harmonic oscillator possesses natural scales for length and energy, which can be used to simplify the problem. These can be found by nondimensionalization.

The result is that, if we measure *energy* in units of $\hbar\omega$ and *distance* in units of $\sqrt{\hbar}/(m\omega)$, then the Hamiltonian simplifies to

$$H = -\tfrac{1}{2}\frac{d^2}{dx^2} + \tfrac{1}{2}x^2,$$

while the energy eigenfunctions and eigenvalues simplify to

$$\psi_n(x) \equiv \langle x \mid n \rangle = \frac{1}{\sqrt{2^n n!}} \pi^{-1/4} \exp(-x^2/2) H_n(x),$$

$$E_n = n + \tfrac{1}{2},$$

where $H_n(x)$ are the Hermite polynomials.

To avoid confusion, we will not adopt these "natural units" in this article. However, they frequently come in handy when performing calculations, by bypassing clutter.

For example, the fundamental solution (propagator) of $H - i\partial_t$, the time-dependent Schrödinger operator for this oscillator, simply boils down to the Mehler kernel,

$$\langle x \mid \exp(-itH) \mid y \rangle \equiv K(x,y;t) = \frac{1}{\sqrt{2\pi i \sin t}} \exp\left(\frac{i}{2\sin t}\left((x^2+y^2)\cos t - 2xy\right)\right),$$

where $K(x,y;0) = \delta(x-y)$. The most general solution for a given initial configuration $\psi(x,0)$ then is simply

$$\psi(x,t) = \int dy\, K(x,y;t)\psi(y,0).$$

Phase Space Solutions

In the phase space formulation of quantum mechanics, solutions to the quantum harmonic oscillator in several different representations of the quasiprobability distribution can be written in closed form. The most widely used of these is for the Wigner quasiprobability distribution, which has the solution

$$F_n(u) = \frac{(-1)^n}{\pi\hbar} L_n\left(4\frac{u}{\hbar\omega}\right)e^{-2u/\hbar\omega},$$

where

$$u = \frac{1}{2}m\omega^2 x^2 + \frac{p^2}{2m}$$

and L_n are the Laguerre polynomials.

This example illustrates how the Hermite and Laguerre polynomials are linked through the Wigner map.

N-dimensional Harmonic Oscillator

The one-dimensional harmonic oscillator is readily generalizable to N dimensions, where $N = 1, 2, 3, \ldots$. In one dimension, the position of the particle was specified by a single coordinate, x. In N dimensions, this is replaced by N position coordinates, which we label $x_1, \ldots, x_N$. Corresponding to each position coordinate is a momentum; we label these $p_1, \ldots, p_N$. The canonical commutation relations between these operators are

$$[x_i, p_j] = i\hbar\delta_{i,j}$$
$$[x_i, x_j] = 0$$
$$[p_i, p_j] = 0$$

The Hamiltonian for this system is

$$H = \sum_{i=1}^{N}\left(\frac{p_i^2}{2m} + \frac{1}{2}m\omega^2 x_i^2\right).$$

As the form of this Hamiltonian makes clear, the N-dimensional harmonic oscillator is exactly analogous to N independent one-dimensional harmonic oscillators with the same mass and spring constant. In this case, the quantities $x_1, ..., x_N$ would refer to the positions of each of the N particles. This is a convenient property of the r^2 potential, which allows the potential energy to be separated into terms depending on one coordinate each.

This observation makes the solution straightforward. For a particular set of quantum numbers $\{n\}$ the energy eigenfunctions for the N-dimensional oscillator are expressed in terms of the 1-dimensional eigenfunctions as:

$$\langle \mathbf{x} | \psi_{\{n\}} \rangle = \prod_{i=1}^{N} \langle x_i | \psi_{n_i} \rangle$$

In the ladder operator method, we define N sets of ladder operators,

$$a_i = \sqrt{\frac{m\omega}{2\hbar}} \left(x_i + \frac{i}{m\omega} p_i \right),$$

$$\backslash a_i^\dagger = \sqrt{\frac{m\omega}{2\hbar}} \left(x_i - \frac{i}{m\omega} p_i \right).$$

By an analogous procedure to the one-dimensional case, we can then show that each of the a_i and $a_i^\dagger$ operators lower and raise the energy by $\hbar\omega$ respectively. The Hamiltonian is

$$H = \hbar\omega \sum_{i=1}^{N} \left(a_i^\dagger a_i + \frac{1}{2} \right).$$

This Hamiltonian is invariant under the dynamic symmetry group $U(N)$ (the unitary group in N dimensions), defined by

$$U a_i^\dagger U^\dagger = \sum_{j=1}^{N} a_j^\dagger U_{ji} \quad \text{for all} \quad U \in U(N),$$

where U_{ji} is an element in the defining matrix representation of $U(N)$.

The energy levels of the system are

$$E = \hbar\omega \left[(n_1 + \cdots + n_N) + \frac{N}{2} \right].$$

$n_i = 0, 1, 2, \ldots$ (the energy level in dimension i).

As in the one-dimensional case, the energy is quantized. The ground state energy is N times the one-dimensional energy, as we would expect using the analogy to N independent one-dimensional oscillators. There is one further difference: in the one-dimensional case, each energy level corresponds to a unique quantum state. In N-dimensions, except for the ground state, the energy levels are *degenerate*, meaning there are several states with the same energy.

The degeneracy can be calculated relatively easily. As an example, consider the 3-dimensional case: Define $n = n_1 + n_2 + n_3$. All states with the same n will have the same energy. For a given n, we choose a particular n_1. Then $n_2 + n_3 = n - n_1$. There are $n - n_1 + 1$ possible pairs $\{n_2, n_3\}$. n_2 can take on the values 0 to $n - n_1$, and for each n_2 the value of n_3 is fixed. The degree of degeneracy therefore is:

$$g_n = \sum_{n_1=0}^{n} n - n_1 + 1 = \frac{(n+1)(n+2)}{2}$$

Formula for general N and n [g_n being the dimension of the symmetric irreducible n^{th} power representation of the unitary group $U(N)$]:

$$g_n = \binom{N+n-1}{n}$$

The special case $N = 3$, given above, follows directly from this general equation. This is however, only true for distinguishable particles, or one particle in N dimensions (as dimensions are distinguishable). For the case of N bosons in a one-dimension harmonic trap, the degeneracy scales as the number of ways to partition an integer n using integers less than or equal to N.

$$g_n = p(N_-, n)$$

This arises due to the constraint of putting N quanta into a state ket where $\sum_{k=0}^{\infty} kn_k = n$ and $\sum_{k=0}^{\infty} n_k = N$, which are the same constraints as in integer partition.

Example: 3D Isotropic Harmonic Oscillator

The Schrödinger equation of a spherically-symmetric three-dimensional harmonic oscillator can be solved explicitly by separation of variables; see this article for the present case. This procedure is analogous to the separation performed in the hydrogen-like atom problem, but with the spherically symmetric potential

$$V(r) = \frac{1}{2}\mu\omega^2 r^2,$$

where μ is the mass of the problem. (Because m will be used below for the magnetic quantum number, mass is indicated by μ, instead of m, as earlier in this article.)

The solution reads

$$\psi_{klm}(r,\theta,\phi) = N_{kl} r^l e^{-vr^2} L_k^{(l+\frac{1}{2})}(2vr^2) Y_{lm}(\theta,\phi)$$

where

$$N_{kl} = \sqrt{\sqrt{\frac{2v^3}{\pi}} \frac{2^{k+2l+3} \, k! v^l}{(2k+2l+1)!!}}$$ is a normalization constant; $v \equiv \dfrac{\mu\omega}{2\hbar}$;

$$L_k^{(l+\frac{1}{2})}(2vr^2)$$

are generalized Laguerre polynomials; The order k of the polynomial is a non-negative integer;

$Y_{lm}(\theta,\phi)$ is a spherical harmonic function;

$\hbar$ is the reduced Planck constant: $\hbar \equiv \dfrac{h}{2\pi}$.

The energy eigenvalue is

$$E = \hbar\omega\left(2k+l+\frac{3}{2}\right).$$

The energy is usually described by the single quantum number

$$n \equiv 2k+l \ .$$

Because k is a non-negative integer, for every even n we have l = 0, 2, ...,n – 2, n and for every odd n we have l =1,3,...,n – 2,n . The magnetic quantum number m is an integer satisfying $-l \le m \le l$, so for every n and l there are $2l$ + 1 different quantum states, labeled by m . Thus, the degeneracy at level n is

$$\sum_{l=...,n-2,n} (2l+1) = \frac{(n+1)(n+2)}{2} \ ,$$

where the sum starts from 0 or 1, according to whether n is even or odd. This result is in accordance with the dimension formula above, and amounts to the dimensionality of a symmetric representation of SU(3), the relevant degeneracy group.

Harmonic Oscillators Lattice: Phonons

We can extend the notion of a harmonic oscillator to a one lattice of many particles. Consider a one-dimensional quantum mechanical *harmonic chain* of N identical atoms. This is the simplest quantum mechanical model of a lattice, and we will see how phonons arise from it. The formalism that we will develop for this model is readily generalizable to two and three dimensions.

As in the previous section, we denote the positions of the masses by x_1, x_2,..., as measured from their equilibrium positions (i.e. x_i = 0 if the particle i is at its equilibrium position.) In two or more dimensions, the x_i are vector quantities. The Hamiltonian for this system is

$$\mathbf{H} = \sum_{i=1}^{N} \frac{p_i^2}{2m} + \frac{1}{2}m\omega^2 \sum_{\{ij\}(nn)} (x_i - x_j)^2 \ ,$$

where m is the (assumed uniform) mass of each atom, and x_i and p_i are the position and momentum operators for the i th atom and the sum is made over the nearest neighbors (nn). However, it is customary to rewrite the Hamiltonian in terms of the normal modes of the wavevector rather than in terms of the particle coordinates so that one can work in the more convenient Fourier space.

We introduce, then, a set of N "normal coordinates" Q_k, defined as the discrete Fourier transforms of the xs, and N "conjugate momenta" Π defined as the Fourier transforms of the ps,

$$Q_k \quad \frac{1}{\sqrt{N}}\sum_l e^{ikal}x_l$$

$$\Pi_k = \frac{1}{\sqrt{N}}\sum_l e^{-ikal}p_l.$$

The quantity k_n will turn out to be the wave number of the phonon, i.e. 2π divided by the wavelength. It takes on quantized values, because the number of atoms is finite.

This preserves the desired commutation relations in either real space or wave vector space

$$\left[x_l,p_m\right]=i\hbar\delta_{l,m}\left[Q_k,\Pi_{k'}\right]=\frac{1}{N}\sum_{l,m}e^{ikal}e^{-ik'am}\left[x_l,p_m\right]=\frac{i\hbar}{N}\sum_m e^{iam(k-k')}=i\hbar\delta_{k,k'}\left[Q_k,Q_{k'}\right]=\left[\Pi_k,\Pi_{k'}\right]=0\,.$$

From the general result

$$\sum_l x_l x_{l+m}=\frac{1}{N}\sum_{kk'}Q_k Q_{k'}\sum_l e^{ial(k+k')}e^{iamk'}=\sum_k Q_k Q_{-k}e^{iamk}$$

$$\sum_l p_l^{\,2}=\sum_k \Pi_k \Pi_{-k},$$

it is easy to show, through elementary trigonometry, that the potential energy term is

$$\frac{1}{2}m\omega^2\sum_j (x_j-x_{j+1})^2=\frac{1}{2}m\omega^2\sum_k Q_k Q_{-k}(2-e^{ika}-e^{-ika})=\frac{1}{2}m\sum_k \omega_k^{\,2}Q_k Q_{-k},$$

where

$$\omega_k=\sqrt{2\omega^2(1-\cos(ka))}\,.$$

The Hamiltonian may be written in wave vector space as

$$\mathbf{H}=\frac{1}{2m}\sum_k \left(\Pi_k\Pi_{-k}+m^2\omega_k^2 Q_k Q_{-k}\right).$$

Note that the couplings between the position variables have been transformed away;

if the Qs and Πs were hermitian(which they are not), the transformed Hamiltonian would describe N *uncoupled* harmonic oscillators.

The form of the quantization depends on the choice of boundary conditions; for simplicity, we impose *periodic* boundary conditions, defining the $(N + 1)$th atom as equivalent to the irst atom. Physically, this corresponds to joining the chain at its ends. The resulting quantization is

$$k = k_n = \frac{2n\pi}{Na} \quad \text{for } n = 0, \pm 1, \pm 2, \ldots, \pm \frac{N}{2}.$$

The upper bound to n comes from the minimum wavelength, which is twice the lattice spacing a, as discussed above.

The harmonic oscillator eigenvalues or energy levels for the mode ω_k are

$$E_n = \left(\frac{1}{2} + n \right) \hbar \omega_k \qquad n = 0, 1, 2, 3, \ldots$$

If we ignore the zero-point energy then the levels are evenly spaced at

$$\hbar\omega, 2\hbar\omega, 3\hbar\omega, \ldots$$

So an exact amount of energy $\hbar\omega$, must be supplied to the harmonic oscillator lattice to push it to the next energy level. In comparison to the photon case when the electromagnetic field is quantised, the quantum of vibrational energy is called a phonon.

All quantum systems show wave-like and particle-like properties. The particle-like properties of the phonon are best understood using the methods of second quantization and operator techniques described later.

Applications

- The vibrations of a diatomic molecule are an example of a two-body version of the quantum harmonic oscillator. In this case, the angular frequency is given by

$$\omega = \sqrt{\frac{k}{\mu}}$$

  where $\mu = m_1 m_2 / (m_1 + m_2)$ is the reduced mass and is determined by the masses m_1, m_2 of the two atoms.

- The Hooke's atom is a simple model of the helium atom using the quantum harmonic oscillator

- Modelling phonons, as discussed above

- A charge, q, with mass, m, in a uniform magnetic field, $\boldsymbol{B}$, is an example of a one-dimensional quantum harmonic oscillator: the Landau quantization.

Franck–Condon Principle

The Franck–Condon principle is a rule in spectroscopy and quantum chemistry that explains the intensity of vibronic transitions. Vibronic transitions are the simultaneous changes in electronic and vibrational energy levels of a molecule due to the absorption or emission of a photon of the appropriate energy. The principle states that during an electronic transition, a change from one vibrational energy level to another will be more likely to happen if the two vibrational wave functions overlap more significantly.

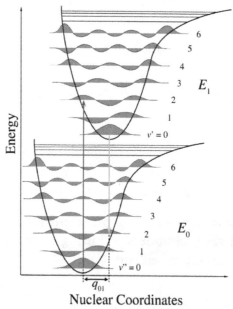

Figure 1. Franck–Condon principle energy diagram. Since electronic transitions are very fast compared with nuclear motions, vibrational levels are favored when they correspond to a minimal change in the nuclear coordinates. The potential wells are shown favoring transitions between $v = 0$ and $v = 2$.

Overview

The Franck–Condon principle has a well-established semiclassical interpretation based on the original contributions of James Franck [Franck 1926]. Electronic transitions are essentially instantaneous compared with the time scale of nuclear motions, therefore if the molecule is to move to a new vibrational level during the electronic transition, this new vibrational level must be instantaneously compatible with the nuclear positions and momenta of the vibrational level of the molecule in the originating electronic state. In the semiclassical picture of vibrations (oscillations) of a simple harmonic oscillator, the necessary conditions can occur at the turning points, where the momentum is zero.

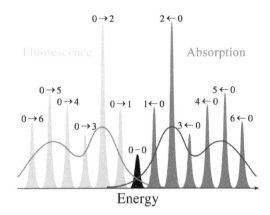

Figure 2. Schematic representation of the absorption and fluorescence spectra corresponding to the energy diagram in Figure 1. The symmetry is due to the equal shape of the ground and excited state potential wells. The narrow lines can usually only be observed in the spectra of dilute gases. The darker curves represent the inhomogeneous broadening of the same transitions as occurs in liquids and solids. Electronic transitions between the lowest vibrational levels of the electronic states (the 0–0 transition) have the same energy in both absorption and fluorescence.

Classically, the Franck–Condon principle is the approximation that an electronic transition is most likely to occur without changes in the positions of the nuclei in the molecular entity and its environment. The resulting state is called a Franck–Condon state, and the transition involved, a vertical transition. The quantum mechanical formulation of this principle is that the intensity of a vibronic transition is proportional to the square of the overlap integral between the vibrational wavefunctions of the two states that are involved in the transition.

— IUPAC Compendium of Chemical Terminology, 2nd Edition (1997)

In the quantum mechanical picture, the vibrational levels and vibrational wavefunctions are those of quantum harmonic oscillators, or of more complex approximations to the potential energy of molecules, such as the Morse potential. Figure 1 illustrates the Franck–Condon principle for vibronic transitions in a molecule with Morse-like potential energy functions in both the ground and excited electronic states. In the low temperature approximation, the molecule starts out in the $v = 0$ vibrational level of the ground electronic state and upon absorbing a photon of the necessary energy, makes a transition to the excited electronic state. The electron configuration of the new state may result in a shift of the equilibrium position of the nuclei constituting the molecule. In the figure this shift in nuclear coordinates between the ground and the first excited state is labeled as q 01. In the simplest case of a diatomic molecule the nuclear coordinates axis refers to the internuclear separation. The vibronic transition is indicated by a vertical arrow due to the assumption of constant nuclear coordinates during the transition. The probability that the molecule can end up in any particular vibrational level is proportional to the square of the (vertical) overlap of the vibrational wavefunctions of the original and final state (Quantum mechanical formulation section is below). In the electronic excited state molecules quickly relax to the lowest vibrational level of the lowest electronic excitation state (Kasha's rule), and from there can decay to the electronic ground state via photon emission. The Franck–Condon principle is applied equally to absorption and to fluorescence.

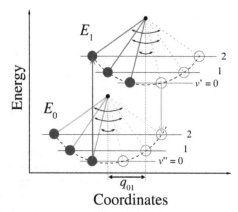

Figure 3. Semiclassical pendulum analogy of the Franck-Condon principle. Vibronic transitions are allowed at the classical turning points because both the momentum and the nuclear coordinates correspond in the two represented energy levels. In this illustration, the 0−2 vibrational transitions are favored.

The applicability of the Franck–Condon principle in both absorption and fluorescence, along with Kasha's rule leads to an approximate mirror symmetry shown in Figure 2. The vibrational structure of molecules in a cold, sparse gas is most clearly visible due to the absence of inhomogeneous broadening of the individual transitions. Vibronic transitions are drawn in Figure 2 as narrow, equally spaced Lorentzian lineshapes. Equal spacing between vibrational levels is only the case for the parabolic potential of simple harmonic oscillators, in more realistic potentials, such as those shown in Figure 1, energy spacing decreases with increasing vibrational energy. Electronic transitions to and from the lowest vibrational states are often referred to as 0–0 (zero zero) transitions and have the same energy in both absorption and fluorescence.

Development of the Principle

In a report published in 1926 in Transactions of the Faraday Society, James Franck was concerned with the mechanisms of photon-induced chemical reactions. The presumed mechanism was the excitation of a molecule by a photon, followed by a collision with another molecule during the short period of excitation. The question was whether it was possible for a molecule to break into photoproducts in a single step, the absorption of a photon, and without a collision. In order for a molecule to break apart, it must acquire from the photon a vibrational energy exceeding the dissociation energy, that is, the energy to break a chemical bond. However, as was known at the time, molecules will only absorb energy corresponding to allowed quantum transitions, and there are no vibrational levels above the dissociation energy level of the potential well. High-energy photon absorption leads to a transition to a higher electronic state instead of dissociation. In examining how much vibrational energy a molecule could acquire when it is excited to a higher electronic level, and whether this vibrational energy could be enough to immediately break apart the molecule, he drew three diagrams representing the possible changes in binding energy between the lowest electronic state and higher electronic states.

Diagram I. shows a great weakening of the binding on a transition from the normal state n to the excited states a and a'. Here we have D > D' and D' > D". At the same time the equilibrium position of the nuclei moves with the excitation to greater values of r. If we go from the equilibrium position (the minimum of potential energy) of the n curve vertically [emphasis added] upwards to the a curves in Diagram I. the particles will have a potential energy greater than D' and will fly apart. In this case we have a very great change in the oscillation energy on excitation by light...

—James Franck, 1926

James Franck recognized that changes in vibrational levels could be a consequence of the instantaneous nature of excitation to higher electronic energy levels and a new equilibrium position for the nuclear interaction potential. Edward Condon extended this insight beyond photoreactions in a 1926 Physical Review article titled "A Theory of Intensity Distribution in Band Systems". Here he formulates the semiclassical formulation in a manner quite similar to its modern form. The first joint reference to both Franck and Condon in regards to the new principle appears in the same 1926 issue of Physical Review in an article on the band structure of carbon monoxide by Raymond Birge.

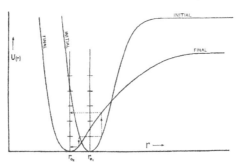

Fig. 1. Typical relation of the potential energy curves, illustrating graphical method of finding favored transitions.

Figure 5. Figure 1 in Edward Condon's first publication on what is now the Franck–Condon principle [Condon 1926]. Condon chose to superimpose the potential curves to illustrate the method of estimating vibrational transitions.

Quantum Mechanical Formulation

Consider an electrical dipole transition from the initial vibrational state (v) of the ground electronic level (ε), $| \varepsilon v \rangle$, to some vibrational state (v') of an excited electronic state (ε'), $| \varepsilon' v' \rangle$. The molecular dipole operator μ is determined by the charge ($-e$) and locations ($\mathbf{r}_i$) of the electrons as well as the charges ($+Z_j e$) and locations ($\mathbf{R}_j$) of the nuclei:

$$\mu = \mu_e + \mu_N = -e\sum_i \mathbf{r}_i + e\sum_j Z_j \mathbf{R}_j.$$

The probability amplitude P for the transition between these two states is given by

$$P = \langle \psi' | \mu | \psi \rangle = \int \psi'^* \mu \psi \, d\tau,$$

where ψ and ψ' are, respectively, the overall wavefunctions of the initial and final state. The overall wavefunctions are the product of the individual vibrational (depending on spatial coordinates of the nuclei) and electronic space and spin wavefunctions:

$$\psi = \psi_e \psi_v \psi_s.$$

This separation of the electronic and vibrational wavefunctions is an expression of the Born–Oppenheimer approximation and is the fundamental assumption of the Franck–Condon principle. Combining these equations leads to an expression for the probability amplitude in terms of separate electronic space, spin and vibrational contributions:

$$P = \langle \psi_{e'} \psi_{v'} \psi_{s'} | \mu | \psi_e \psi_v \psi_s \rangle = \int \psi_{e'}^* \psi_{v'}^* \psi_{s'}^* (\mu_e + \mu_N) \psi_e \psi_v \psi_s \, d\tau$$

$$= \int \psi_{e'}^* \psi_{v'}^* \psi_{s'}^* \mu_e \psi_e \psi_v \psi_s \, d\tau + \int \psi_{e'}^* \psi_{v'}^* \psi_{s'}^* \mu_N \psi_e \psi_v \psi_s \, d\tau$$

$$= \underbrace{\int \psi_{v'}^* \psi_v \, d\tau_n}_{\substack{\text{Franck–Condon} \\ \text{factor}}} \underbrace{\int \psi_{e'}^* \mu_e \psi_e \, d\tau_e}_{\substack{\text{orbital} \\ \text{selection rule}}} \underbrace{\int \psi_{s'}^* \psi_s \, d\tau_s}_{\substack{\text{spin} \\ \text{selection rule}}} + \int \psi_{e'}^* \psi_e \, d\tau_e \underbrace{\int \psi_{v'}^* \mu_N \psi_v \, d\tau_v}_{0} \int \psi_{s'}^* \psi_s \, d\tau_s.$$

The spin-independent part of the initial integral is here *approximated* as a product of two integrals:

$$\iint \psi_{v'}^* \psi_{e'}^* \mu_e \psi_e \psi_v \, d\tau_e d\tau_n \approx \int \psi_{v'}^* \psi_v \, d\tau_n \int \psi_{e'}^* \mu_e \psi_e \, d\tau_e.$$

This factorization would be exact if the integral $\int \psi_{e'}^* \mu_e \psi_e \, d\tau_e$ over the spatial coordinates of the electrons would not depend on the nuclear coordinates. However, in the Born–Oppenheimer approximation ψ_e and ψ_e' do depend (parametrically) on the nuclear coordinates, so that the integral (a so-called *transition dipole surface*) is a function of nuclear coordinates. Since the dependence is usually rather smooth it is neglected (i.e., the assumption that the transition dipole surface is independent of nuclear coordinates, called the *Condon approximation* is often allowed).

The first integral after the plus sign is equal to zero because electronic wavefunctions of different states are orthogonal. Remaining is the product of three integrals. The first integral is the vibrational overlap integral, also called the Franck-Condon factor. The remaining two integrals contributing to the probability amplitude determine the electronic spatial and spin selection rules.

The Franck–Condon principle is a statement on allowed vibrational transitions between two *different* electronic states; other quantum mechanical selection rules may lower the probability of a transition or prohibit it altogether. Rotational selection rules have been neglected in the above derivation. Rotational contributions can be observed in the spectra of gases but are strongly suppressed in liquids and solids.

It should be clear that the quantum mechanical formulation of the Franck–Condon principle is the result of a series of approximations, principally the electrical dipole transition assumption and the Born–Oppenheimer approximation. Weaker magnetic dipole and electric quadrupole electronic transitions along with the incomplete validity of the factorization of the total wavefunction into nuclear, electronic spatial and spin wavefunctions means that the selection rules, including the Franck-Condon factor, are not strictly observed. For any given transition, the value of P is determined by all of the selection rules, however spin selection is the largest contributor, followed by electronic selection rules. The Franck-Condon factor only *weakly* modulates the intensity of transitions, i.e., it contributes with a factor on the order of 1 to the intensity of bands whose order of magnitude is determined by the other selection rules. The table below gives the range of extinction coefficients for the possible combinations of allowed and forbidden spin and orbital selection rules.

Intensities of Electronic Transitions	
	Range of extinction coefficient (ε) values (mol^{-1} cm^{-1})
Spin and orbitally allowed	10^3 to 10^5
Spin allowed but orbitally forbidden	10^0 to 10^3
Spin forbidden but orbitally allowed	10^{-5} to 10^0

Franck–Condon Metaphors in Spectroscopy

The Franck–Condon principle, in its canonical form, applies only to changes in the vibrational levels of a molecule in the course of a change in electronic levels by either absorption or emission of a photon. The physical intuition of this principle is anchored by the idea that the nuclear coordinates of the atoms constituting the molecule do not have time to change during the very brief amount of time involved in an electronic transition. However, this physical intuition can be, and is indeed, routinely extended to interactions between light-absorbing or emitting molecules (chromophores) and their environment. Franck–Condon metaphors are appropriate because molecules often interact strongly with surrounding molecules, particularly in liquids and solids, and these interactions modify the nuclear coordinates of the chromophore in ways closely analogous to the molecular vibrations considered by the Franck–Condon principle.

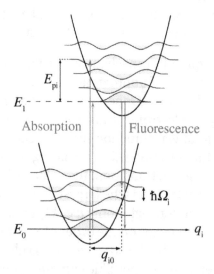

Figure 6. Energy diagram of an electronic transition with phonon coupling along the configurational coordinate q_i, a normal mode of the lattice. The upwards arrows represent absorption without phonons and with four phonons. The downwards arrows represent the symmetric process in emission.

Franck–Condon Principle for Phonons

The closest Franck–Condon analogy is due to the interaction of phonons (quanta of lattice vibrations) with the electronic transitions of chromophores embedded as impurities in the lattice. In this situation, transitions to higher electronic levels can take place when the energy of the photon corresponds to the purely electronic transition energy or to the purely electronic transition energy plus the energy of one or more lattice phonons. In the low-temperature approximation, emission is from the zero-phonon level of the excited state to the zero-phonon level of the ground state or to higher phonon levels of the ground state. Just like in the Franck–Condon principle, the probability of transitions involving phonons is determined by the overlap of the phonon wavefunctions at the initial and final energy levels. For the Franck–Condon principle applied to phonon transitions, the label of the horizontal axis of Figure 1 is replaced in Figure 6 with the configurational coordinate for a normal mode. The lattice mode q_i potential energy in Figure 6 is represented as that of a harmonic oscillator, and the spacing between phonon levels ($\hbar\Omega_i$) is determined by lattice parameters. Because the energy of single phonons is generally quite small, zero- or few-phonon transitions can only be observed at temperatures below about 40 kelvins.

Franck–Condon Principle in Solvation

Franck–Condon considerations can also be applied to the electronic transitions of chromophores dissolved in liquids. In this use of the Franck–Condon metaphor, the vibrational levels of the chromophores, as well as interactions of the chromophores with phonons in the liquid, continue to contribute to the structure of the absorption and emission spectra, but these effects are considered separately and independently.

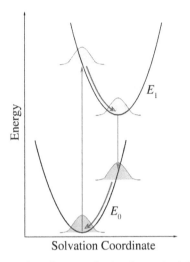

Figure 7. Energy diagram illustrating the Franck–Condon principle applied to the solvation of chromophores. The parabolic potential curves symbolize the interaction energy between the chromophores and the solvent. The Gaussian curves represent the distribution of this interaction energy.

Consider chromophores surrounded by solvent molecules. These surrounding molecules may interact with the chromophores, particularly if the solvent molecules are polar. This association between solvent and solute is referred to as solvation and is a stabilizing interaction, that is, the solvent molecules can move and rotate until the energy of the interaction is minimized. The interaction itself involves electrostatic and van der Waals forces and can also include hydrogen bonds. Franck–Condon principles can be applied when the interactions between the chromophore and the surrounding solvent molecules are different in the ground and in the excited electronic state. This change in interaction can originate, for example, due to different dipole moments in these two states. If the chromophore starts in its ground state and is close to equilibrium with the surrounding solvent molecules and then absorbs a photon that takes it to the excited state, its interaction with the solvent will be far from equilibrium in the excited state. This effect is analogous to the original Franck–Condon principle: the electronic transition is very fast compared with the motion of nuclei—the rearrangement of solvent molecules in the case of solvation. We now speak of a vertical transition, but now the horizontal coordinate is solvent-solute interaction space. This coordinate axis is often labeled as "Solvation Coordinate" and represents, somewhat abstractly, all of the relevant dimensions of motion of all of the interacting solvent molecules.

In the original Franck–Condon principle, after the electronic transition, the molecules which end up in higher vibrational states immediately begin to relax to the lowest vibrational state. In the case of solvation, the solvent molecules will immediately try to rearrange themselves in order to minimize the interaction energy. The rate of solvent relaxation depends on the viscosity of the solvent. Assuming the solvent relaxation time is short compared with the lifetime of the electronic excited state, emission will be from the lowest solvent energy state of the excited electronic state. For small-molecule solvents such as water or methanol at ambient temperature, solvent relaxation time is on the order of some tens of picoseconds whereas chromophore excited state lifetimes range from a few picoseconds to a few nanoseconds. Immediately after the transition to the ground electronic state, the solvent molecules must also rearrange themselves to accommodate the new electronic configuration of the chromophore. Figure 7 illustrates the Franck–Condon principle applied to solvation. When the solution is illuminated by with light corresponding to the electronic transition energy, some of the chromophores will move to the excited state. Within this group of chromophores there will be a statistical distribution of solvent-chromophore interaction energies, represented in the figure by a Gaussian distribution function. The solvent-chromophore interaction is drawn as a parabolic potential in both electronic states. Since the electronic transition is essentially instantaneous on the time scale of solvent motion (vertical arrow), the collection of excited state chromophores is immediately far from equilibrium. The rearrangement of the solvent molecules according to the new potential energy curve is represented by the curved arrows in Figure 7. Note that while the electronic transitions are quantized, the chromophore-solvent interaction energy is treated as a classical continuum due to the large number of molecules involved. Although emission is depicted as taking place from the minimum of the excited state chromophore-solvent interaction potential, significant emission can take place before equilibrium is reached when the viscosity of the solvent is high or the lifetime of the excited state is short. The energy difference between absorbed and emitted photons depicted in Figure 7 is the solvation contribution to the Stokes shift.

Born–Oppenheimer Approximation

In quantum chemistry and molecular physics, the Born–Oppenheimer (BO) approximation is the assumption that the motion of atomic nuclei and electrons in a molecule can be separated. The approach is named after Max Born and J. Robert Oppenheimer. In mathematical terms, it allows the wavefunction of a molecule to be broken into its electronic and nuclear (vibrational, rotational) components.

$$\Psi_{total} = \psi_{electronic} \times \psi_{nuclear}$$

Computation of the energy and the wavefunction of an average-size molecule is simplified by the approximation. For example, the benzene molecule consists of 12 nuclei and 42 electrons. The time independent Schrödinger equation, which must be solved to obtain the energy and wavefunction of this molecule, is a partial differential eigenvalue equation in 162 variables—the spatial coordinates of *the electrons and the nuclei*. The BO approximation makes it possible to compute the wavefunction in two less complicated consecutive steps. This approximation was proposed in 1927, in the early period of quantum mechanics, by Born and Oppenheimer and is still indispensable in quantum chemistry.

In the first step of the BO approximation the *electronic* Schrödinger equation is solved, yielding the wavefunction $\Psi_{\text{electronic}}$ depending on electrons only. For benzene this wavefunction depends on 126 electronic coordinates. During this solution the nuclei are fixed in a certain configuration, very often the equilibrium configuration. If the effects of the quantum mechanical nuclear motion are to be studied, for instance because a vibrational spectrum is required, this electronic computation must be in nuclear coordinates. In the second step of the BO approximation this function serves as a potential in a Schrödinger equation *containing only the nuclei*—for benzene an equation in 36 variables.

The success of the BO approximation is due to the difference between nuclear and electronic masses. The approximation is an important tool of quantum chemistry; without it only the lightest molecule, H_2, could be handled, and all computations of molecular wavefunctions for larger molecules make use of it. Even in the cases where the BO approximation breaks down, it is used as a point of departure for the computations.

The electronic energies consist of kinetic energies, interelectronic repulsions, internuclear repulsions, and electron–nuclear attractions. In accord with the Hellmann-Feynman theorem, the nuclear potential is taken to be an average over electron configurations of the sum of the electron–nuclear and internuclear electric potentials.

In molecular spectroscopy, because the ratios of the periods of the electronic, vibrational and rotational energies are each related to each other on scales in the order of a thousand, the Born–Oppenheimer name has also been attached to the approximation where the energy components are treated separately.

$$E_{\text{total}} = E_{\text{electronic}} + E_{\text{vibrational}} + E_{\text{rotational}} + E_{\text{nuclear}}$$

The nuclear spin energy is so small that it is normally omitted.

Short Description

The Born–Oppenheimer (BO) approximation is ubiquitous in quantum chemical calculations of molecular wavefunctions. It consists of two steps.

In the first step the nuclear kinetic energy is neglected, that is, the corresponding operator T_n is subtracted from the total molecular Hamiltonian. In the remaining electronic Hamiltonian H_e the nuclear positions enter as parameters. The electron–nucleus interactions are *not* removed and the electrons still "feel" the Coulomb potential of the nuclei clamped at certain positions in space. (This first step of the BO approximation is therefore often referred to as the *clamped nuclei* approximation.)

The electronic Schrödinger equation

$$H_e(\mathbf{r}, \mathbf{R})\, \chi(\mathbf{r}, \mathbf{R}) = E_e\, \chi(\mathbf{r}, \mathbf{R})$$

is solved (out of necessity, approximately). The quantity $\mathbf{r}$ stands for all electronic coordinates and R for all nuclear coordinates. The electronic energy eigenvalue E_e depends on the chosen positions R of the nuclei. Varying these positions R in small steps and repeatedly solving the electronic Schrödinger equation, one obtains E_e as a function of R. This is the potential energy surface (PES): $E_e(\mathrm{R})$. Because this procedure of recomputing the electronic wave functions as a function of an infinitesimally changing nuclear geometry is reminiscent of the conditions for the adiabatic theorem, this manner of obtaining a PES is often referred to as the *adiabatic approximation* and the PES itself is called an *adiabatic surface*.

In the second step of the BO approximation the nuclear kinetic energy T_n (containing partial derivatives with respect to the components of R) is reintroduced and the Schrödinger equation for the nuclear motion

$$\left[T_n + E_e(\mathbf{R})\right]\phi(\mathbf{R}) = E\phi(\mathbf{R})$$

is solved. This second step of the BO approximation involves separation of vibrational, translational, and rotational motions. This can be achieved by application of the Eckart conditions. The eigenvalue E is the total energy of the molecule, including contributions from electrons, nuclear vibrations, and overall rotation and translation of the molecule.

Derivation of the Born–Oppenheimer Approximation

It will be discussed how the BO approximation may be derived and under which conditions it is applicable. At the same time we will show how the BO approximation may be improved by including vibronic coupling. To that end the second step of the BO approximation is generalized to a set of coupled eigenvalue equations depending on nuclear coordinates only. Off-diagonal elements in these equations are shown to be nuclear kinetic energy terms.

It will be shown that the BO approximation can be trusted whenever the PESs, obtained from the solution of the electronic Schrödinger equation, are well separated:

$$E_0(\mathbf{R}) \ll E_1(\mathbf{R}) \ll E_2(\mathbf{R}) \ll \cdots \text{ for all } \mathbf{R}.$$

We start from the *exact* non-relativistic, time-independent molecular Hamiltonian:

$$H = H_e + T_n$$

with

$$H_e = -\sum_i \frac{1}{2}\nabla_i^2 - \sum_{i,A} \frac{Z_A}{r_{iA}} + \sum_{i>j} \frac{1}{r_{ij}} + \sum_{B>A} \frac{Z_A Z_B}{R_{AB}} \quad \text{and} \quad T_n = -\sum_A \frac{1}{2M_A}\nabla_A^2.$$

The position vectors $\mathbf{r} \equiv \{\mathbf{r}_i\}$ of the electrons and the position vectors $\mathbf{R} \equiv \{\mathbf{R}_A = (R_{Axy}, R_{Ayz}, R_{Azx})\}$ of the nuclei are with respect to a Cartesian inertial frame. Distances between particles are written as $r_{iA} \equiv |\mathbf{r}_i - \mathbf{R}_A|$ (distance between electron i and nucleus A) and similar definitions hold for r_{ij} and R_{AB}. We assume that the molecule is in a homogeneous (no external force) and isotropic (no external torque) space. The only interactions are the two-body Coulomb interactions among the electrons and nuclei. The Hamiltonian is expressed in atomic units, so that we do not see Planck's constant, the dielectric constant of the vacuum, electronic charge, or electronic mass in this formula. The only constants explicitly entering the formula are Z_A and M_A—the atomic number and mass of nucleus A.

It is useful to introduce the total nuclear momentum and to rewrite the nuclear kinetic energy operator as follows:

$$T_n = \sum_A \sum_{\alpha=x,y,z} \frac{P_{A\alpha}P_{A\alpha}}{2M_A} \quad \text{with} \quad P_{A\alpha} = -i\frac{\partial}{\partial R_{A\alpha}}.$$

Suppose we have K electronic eigenfunctions $\chi_k(\mathbf{r};\mathbf{R})$ of H_e, that is, we have solved

$$H_e\,\chi_k(\mathbf{r};\mathbf{R}) = E_k(\mathbf{R})\,\chi_k(\mathbf{r};\mathbf{R}) \quad \text{for} \quad k=1,\dots,K.$$

The electronic wave functions χ_k will be taken to be real, which is possible when there are no magnetic or spin interactions. The *parametric dependence* of the functions χ_k on the nuclear coordinates is indicated by the symbol after the semicolon. This indicates that, although χ_k is a real-valued function of $\mathbf{r}$, its functional form depends on $\mathbf{R}$.

For example, in the molecular-orbital-linear-combination-of-atomic-orbitals (LCAO-MO) approximation, χ_k is a molecular orbital (MO) given as a linear expansion of atomic orbitals (AOs). An AO depends visibly on the coordinates of an electron, but the nuclear coordinates are not explicit in the MO. However, upon change of geometry, i.e., change of $\mathbf{R}$, the LCAO coefficients obtain different values and we see corresponding changes in the functional form of the MO χ_k.

We will assume that the parametric dependence is continuous and non-differentiable, so that it is meaningful to consider

$$P_{A\alpha}\chi_k(\mathbf{r};\mathbf{R}) = -i\frac{\partial \chi_k(\mathbf{r};\mathbf{R})}{\partial R_{A\alpha}} \quad \text{for} \quad \alpha = x,y,z,$$

which in general will not be zero.

The total wave function $\Psi(\mathbf{R},\mathbf{r})$ is expanded in terms of $\chi_k(\mathbf{r};\mathbf{R})$:

$$\Psi(\mathbf{R},\mathbf{r}) = \sum_{k=1}^{K} \chi_k(\mathbf{r};\mathbf{R})\phi_k(\mathbf{R}),$$

with

$$\langle \chi_{k'}(\mathbf{r};\mathbf{R})|\chi_k(\mathbf{r};\mathbf{R})\rangle_{(\mathbf{r})} = \delta_{k'k}$$

and where the subscript $(\mathbf{r})$ indicates that the integration, implied by the bra–ket notation, is over electronic coordinates only. By definition, the matrix with general element

$$\left(\mathbb{H}_{\mathrm{e}}(\mathbf{R})\right)_{k'k} \equiv \langle \chi_{k'}(\mathbf{r};\mathbf{R}) \mid H_{\mathrm{e}} \mid \chi_k(\mathbf{r};\mathbf{R})\rangle_{(\mathbf{r})} = \delta_{k'k}E_k(\mathbf{R})$$

is diagonal. After multiplication by the real function $(\mathbf{r};\mathbf{R})$ from the left and integration over the electronic coordinates $\mathbf{r}$ the total Schrödinger equation

$$H\,\Psi(\mathbf{R},\mathbf{r}) = E\,\Psi(\mathbf{R},\mathbf{r})$$

is turned into a set of K coupled eigenvalue equations depending on nuclear coordinates only

$$\left[\mathbb{H}_{\mathrm{n}}(\mathbf{R}) + \mathbb{H}_{\mathrm{e}}(\mathbf{R})\right]\phi(\mathbf{R}) = E\,\phi(\mathbf{R}).$$

The column vector $\phi(\mathbf{R})$ has elements $\phi_k(\mathbf{R})$, $k = 1,\ldots,K$. The matrix $\mathbb{H}_{\mathrm{e}}(\mathbf{R})$ is diagonal and the nuclear Hamilton matrix is non-diagonal with the following off-diagonal (*vibronic coupling*) terms,

$$\left(\mathbb{H}_{\mathrm{n}}(\mathbf{R})\right)_{k'k} = \langle \chi_{k'}(\mathbf{r};\mathbf{R}) \mid T_{\mathrm{n}} \mid \chi_k(\mathbf{r};\mathbf{R})\rangle_{(\mathbf{r})}.$$

The vibronic coupling in this approach is through nuclear kinetic energy terms. Solution of these coupled equations gives an approximation for energy and wavefunction that goes beyond the Born–Oppenheimer approximation. Unfortunately, the off-diagonal kinetic energy terms are usually difficult to handle. This is why often a diabatic transformation is applied, which retains part of the nuclear kinetic energy terms on the diagonal, removes the kinetic energy terms from the off-diagonal and creates coupling terms between the adiabatic PESs on the off-diagonal.

If we can neglect the off-diagonal elements the equations will uncouple and simplify drastically. In order to show when this neglect is justified, we suppress the coordinates in the notation and write, by applying the Leibniz rule for differentiation, the matrix elements of T_{n} as

$$H_{\mathrm{n}}(\mathbf{R})_{k'k} \equiv \left(\mathbb{H}_{\mathrm{n}}(\mathbf{R})\right)_{k'k} = \delta_{k'k}T_{\mathrm{n}} + \sum_{A,\alpha}\frac{1}{M_A}\langle \chi_{k'} \mid \left(P_{A\alpha}\chi_k\right)\rangle_{(\mathbf{r})}P_{A\alpha} + \langle \chi_{k'} \mid \left(T_{\mathrm{n}}\chi_k\right)\rangle_{(\mathbf{r})}.$$

The diagonal ($k' = k$) matrix elements $\langle \chi_k \mid \left(P_{A\alpha}\chi_k\right)\rangle_{(\mathbf{r})}$ of the operator $P_{A\alpha}$ vanish, because we assume time-reversal invariant so χ_k can be chosen to be always real. The off-diagonal matrix elements satisfy

$$\langle \chi_{k'} \mid \left(P_{A\alpha}\chi_k\right)\rangle_{(\mathbf{r})} = \frac{\langle \chi_{k'} \mid \left[P_{A\alpha},H_{\mathrm{e}}\right] \mid \chi_k\rangle_{(\mathbf{r})}}{E_k(\mathbf{R}) - E_{k'}(\mathbf{R})}.$$

The matrix element in the numerator is

$$\langle \chi_{k'} \mid \left[P_{A\alpha},H_{\mathrm{e}}\right] \mid \chi_k\rangle_{(\mathbf{r})} = iZ_A\sum_i \langle \chi_{k'} \mid \frac{(\mathbf{r}_{iA})_\alpha}{r_{iA}^3} \mid \chi_k\rangle_{(\mathbf{r})} \quad \text{with } \mathbf{r}_{iA} \equiv \mathbf{r}_i - \mathbf{R}_A.$$

The matrix element of the one-electron operator appearing on the right hand side is finite. When the two surfaces come close, $E_k(\mathbf{R}) \approx E_{k'}(\mathbf{R})$, the nuclear momentum coupling term becomes large and is no longer negligible. This is the case where the BO approximation breaks down and a coupled set of nuclear motion equations must be considered, instead of the one equation appearing in the second step of the BO approximation.

Conversely, if all surfaces are well separated, all off-diagonal terms can be neglected and hence the whole matrix of P_α^A is effectively zero. The third term on the right hand side of the expression for the matrix element of T_n (the *Born–Oppenheimer diagonal correction*) can approximately be written as the matrix of P_α^A squared and, accordingly, is then negligible also. Only the first (diagonal) kinetic energy term in this equation survives in the case of well-separated surfaces and a diagonal, uncoupled, set of nuclear motion equations results,

$$\left[T_n + E_k(\mathbf{R})\right]\phi_k(\mathbf{R}) = E\phi_k(\mathbf{R}) \quad \text{for} \quad k = 1, \ldots, K,$$

which are the normal second-step of the BO equations discussed above.

We reiterate that when two or more potential energy surfaces approach each other, or even cross, the Born–Oppenheimer approximation breaks down and one must fall back on the coupled equations. Usually one invokes then the diabatic approximation.

The Born–Oppenheimer Approximation with the Correct Symmetry

To include the correct symmetry within the Born–Oppenheimer (BO) approximation, a molecular system presented in terms of (mass-dependent) nuclear coordinates, $\mathbf{q}$, and formed by the two lowest BO adiabatic potential energy surfaces (PES), $u_1(\mathbf{q})$ and $u_2(\mathbf{q})$, is considered. To insure the validity of the BO approximation the energy of the system, E, is assumed to be low enough so that $u_2(\mathbf{q})$ becomes a closed PES in the region of interest, with the exception of sporadic infinitesimal sites surrounding degeneracy points formed by $u_1(\mathbf{q})$ and $u_2(\mathbf{q})$ (designated as (1,2) degeneracy points).

The starting point is the nuclear adiabatic BO (matrix) equation written in the form:

$$\frac{\hbar^2}{2m}(\nabla + \tau)^2 \Psi + (\mathbf{u} - E)\Psi = 0$$

where $\Psi(\mathbf{q})$ is a column vector that contains the unknown nuclear wave functions $\psi_k(\mathbf{q})$, $\mathbf{u}(\mathbf{q})$ is a diagonal matrix which contains the corresponding adiabatic potential energy surfaces $u_k(\mathbf{q})$, m is the reduced mass of the nuclei, E is the total energy of the system, ∇ is the grad operator with respect to the nuclear coordinates $\mathbf{q}$ and $\hat{\mathbf{o}}(\mathbf{q})$ is a matrix which contains the vectorial Non-Adiabatic Coupling Terms (NACT):

$$\hat{\mathbf{o}}_{jk} = \langle \zeta_j | \nabla \zeta_k \rangle$$

Here $|\zeta_n\rangle; n = j, k$ are eigenfunctions of the electronic Hamiltonian assumed to form a complete Hilbert space in the given region in configuration space.

To study the scattering process taking place on the two lowest surfaces one extracts, from the above BO equation, the two corresponding equations:

$$-\frac{\hbar^2}{2m}\nabla^2\psi_1 + (\tilde{u}_1 - E)\psi_1 - \frac{\hbar^2}{2m}[2\tau_{12}\nabla + \nabla\tau_{12}]\psi_2 = 0$$

$$-\frac{\hbar^2}{2m}\nabla^2\psi_2 + (\tilde{u}_2 - E)\psi_2 + \frac{\hbar^2}{2m}[2\tau_{12}\nabla + \nabla\tau_{12}]\psi_1 = 0$$

where $\tilde{u}_k(\mathbf{q}) = u_k(\mathbf{q}) + (\hbar^2/2m)\tau_{12}^2; k = 1,2$, and $\tau_{12}(=\tau_{12}(\mathbf{q}))$ is the (vectorial) NACT responsible for the coupling between $u_1(\mathbf{q})$ and $u_2(\mathbf{q})$.

Next a new function is introduced:

$$\chi = \psi_1 + i\psi_2$$

and the corresponding rearrangements are made:

(i) Multiplying the second equation by i and combining it with the first equation yields the (complex) equation:

$$-\frac{\hbar^2}{2m}\nabla^2\chi + (\tilde{u}_1 - E)\chi + i\frac{\hbar^2}{2m}[2\tau_{12}\nabla + \nabla\tau_{12}]\chi + i(u_1 - u_2)\psi_2 = 0$$

(ii) The last term in this equation can be deleted for the following reasons: At those points where $u_2(\mathbf{q})$ is classically closed, $\psi_2(\mathbf{q}) \sim 0$ by definition, and at those points where $u_2(\mathbf{q})$ becomes classically allowed (which happens at the vicinity of the (1,2) degeneracy points) this implies that: $u_1(\mathbf{q}) \sim u_2(\mathbf{q})$ or $u_1(\mathbf{q}) - u_2(\mathbf{q}) \sim 0$. Consequently the last term is, indeed, negligibly small at every point in the region of interest and the equation simplifies to become:

$$-\frac{\hbar^2}{2m}\nabla^2\chi + (\tilde{u}_1 - E)\chi + i\frac{\hbar^2}{2m}[2\tau_{12}\nabla + \nabla\tau_{12}]\chi = 0$$

In order for this equation to yield a solution with the correct symmetry, it is suggested to apply a perturbation approach based on an elastic potential, $u_0(\mathbf{q})$, which coincides with $u_1(\mathbf{q})$ at the asymptotic region.

The equation with an elastic potential can be solved, in a straightforward manner, by substitution. Thus, if χ_0 is the solution of this equation, it is presented as:

$$\chi_0(q\,|\,\Gamma) = \xi_0(q)exp[-i\int_{\Gamma}dq'\cdot\tau(q'\,|\,\Gamma)]$$

where Γ is an arbitrary contour and the exponential function contains the relevant symmetry as created while moving along Γ.

The function $\xi_0(\mathbf{q})$ can be shown to be a solution of the (unperturbed/elastic) equation:

$$-\frac{\hbar^2}{2m}\nabla^2\xi_0 + (u_0 - E)\xi_0 = 0$$

Having $\chi_0(\mathbf{q}\,|\,\Gamma)$, the full solution of the above decoupled equation takes the form:

$$\chi(\mathbf{q}\,|\,\Gamma) = \chi_0(\mathbf{q}\,|\,\Gamma) + \eta(\mathbf{q}\,|\,\Gamma)$$

where $\eta(\mathbf{q}\,|\,\Gamma)$ satisfies the resulting inhomogeneous equation:

$$-\frac{\hbar^2}{2m}\nabla^2\eta + (\tilde{u}_1 - E)\eta + i\frac{\hbar^2}{2m}[2\tau_{12}\nabla + \nabla\tau_{12}]\eta = (u_1 - u_0)\chi_0$$

In this equation the inhomogeneity ensures the symmetry for the perturbed part of the solution along any contour and therefore for the solution in the required region in configuration space.

The relevance of the present approach was demonstrated while studying a two-arrangement-channel model (containing one inelastic channel and one reactive channel) for which the two adiabatic states were coupled via a Jahn-Teller conical intersection. A nice fit between the symmetry-preserved, single-state, treatment and the corresponding two-state treatment was obtained. This applies in particular to the reactive state-to-state probabilities for which the ordinary BO approximation led to erroneous results, whereas the symmetry-preserving BO approximation produced the accurate results as they followed from solving the two coupled equations.

Jellium

Jellium, also known as the uniform electron gas (UEG) or homogeneous electron gas (HEG), is a quantum mechanical model of interacting electrons in a solid where the positive charges (i.e. atomic nuclei) are assumed to be uniformly distributed in space whence the electron density is a uniform quantity as well in space. This model allows one to focus on the effects in solids that occur due to the quantum nature of electrons and their mutual repulsive interactions (due to like charge) without explicit introduction of the atomic lattice and structure making up a real material. Jellium is often used in solid-state physics as a simple model of delocalized electrons in a metal, where it can qualitatively reproduce features of real metals such as screening, plasmons, Wigner crystallization and Friedel oscillations.

At zero temperature, the properties of jellium depend solely upon the constant electronic density. This lends it to a treatment within density functional theory; the formalism itself provides the basis for the local-density approximation to the exchange-correlation energy density functional.

The term *jellium* was coined by Conyers Herring, alluding to the "positive jelly" background, and the typical metallic behavior it displays.

Hamiltonian

The jellium model treats the electron-electron coupling rigorously. The artificial and structureless background charge interacts electrostatically with itself and the electrons. The jellium Hamiltonian for N-electrons confined within a volume of space Ω, and with electronic density $\rho(\mathbf{r})$ and (constant) background charge density $n(\mathbf{R}) = N/\Omega$ is

$$\hat{H} = \hat{H}_{el} + \hat{H}_{back} + \hat{H}_{el\text{-}back},$$

where

- H_{el} is the electronic Hamiltonian consisting of the kinetic and electron-electron repulsion terms:

$$\hat{H}_{el} = \sum_{i=1}^{N} \frac{p_i^2}{2m} + \sum_{i<j}^{N} \frac{e^2}{|\mathbf{r}_i - \mathbf{r}_j|}$$

- H_{back} is the Hamiltonian of the positive background charge interacting electrostatically with itself:

$$\hat{H}_{back} = \frac{e^2}{2} \int_{\Omega} d\mathbf{R} \int_{\Omega} d\mathbf{R}' \frac{n(\mathbf{R})n(\mathbf{R}')}{|\mathbf{R}-\mathbf{R}'|} = \frac{e^2}{2}\left(\frac{N}{\Omega}\right)^2 \int_{\Omega} d\mathbf{R} \int_{\Omega} d\mathbf{R}' \frac{1}{|\mathbf{R}-\mathbf{R}'|}$$

- $H_{el\text{-}back}$ is the electron-background interaction Hamiltonian, again an electrostatic interaction:

$$\hat{H}_{el\text{-}back} = \int_{\Omega} d\mathbf{r} \int_{\Omega} d\mathbf{R} \frac{\rho(\mathbf{r})n(\mathbf{R})}{|\mathbf{r}-\mathbf{R}|} = -e^2 \frac{N}{\Omega} \sum_{i=1}^{N} \int_{\Omega} d\mathbf{R} \frac{1}{|\mathbf{r}_i - \mathbf{R}|}$$

H_{back} is a constant and, in the limit of an infinite volume, divergent along with $H_{el\text{-}back}$. The divergence is canceled by a term from the electron-electron coupling: the background interactions cancel and the system is dominated by the kinetic energy and coupling of the electrons. Such analysis is done in Fourier space; the interaction terms of the Hamiltonian which remain correspond to the Fourier expansion of the electron coupling for which $\mathbf{q} \neq \mathbf{0}$.

Contributions to the Total Energy

The traditional way to study the electron gas is to start with non-interacting electrons which are governed only by the kinetic energy part of the Hamiltonian, which yields the free electron gas model. The kinetic energy per electron is given by

$$KE = \frac{3}{5}E_F = \frac{3}{5}\frac{\hbar^2 k_F^2}{2m_e} = \frac{2.21}{r_s^2}\text{Ryd}$$

where E_F is the Fermi energy, k_F is the Fermi wave vector, and the last expression shows the dependence on the Wigner-Seitz radius r_s where energy is measured in Rydbergs.

Without doing much work, one can guess that the electron-electron interactions will scale like the inverse of the average electron-electron separation and hence as $1/r_{12}$ (since the Coulomb interaction goes like one over distance between charges) so that if we view the interactions as a small correction to the kinetic energy, we are describing the limit of small r_s (i.e. $1/r_s^2$ being larger than $1/r_s$) and hence high electron density. Unfortunately, real metals typically have between 2-5 which means this picture needs serious revision.

The first correction to the free-electron model for jelium is from the Fock exchange contribution to electron-electron interactions. Adding this in, one has a total energy of

$$E = \frac{2.21}{r_s^2} - \frac{0.916}{r_s}$$

where the negative term is due to exchange: exchange interactions lower the total energy. Higher order corrections to the total energy are due to electron correlation and if one decides to work in a series for small r_s, one finds

$$E = \frac{2.21}{r_s^2} - \frac{0.916}{r_s} + 0.0622\ln(r_s) - 0.096 + O(r_s)$$

The series is quite accurate for small r_s but of dubious value for r_s values found in actual metals.

For the full range of r_s, Chachiyo's correlation energy density can be used as the higher order correction. In this case,

$$E = \frac{2.21}{r_s^2} - \frac{0.916}{r_s} + a\ln\left(1 + \frac{b}{r_s} + \frac{b}{r_s^2}\right)$$

which agrees quite well (on the order of milli-Hartree) with the Quantum Monte Carlo simulation.

Zero-temperature Phase Diagram of Jellium in Three and Two Dimensions

The physics of the zero-temperature phase behavior of jellium is driven by competition between the kinetic energy of the electrons and the electron-electron interaction energy. The kinetic-energy operator in the Hamiltonian scales as $1/r_s^2$, where r_s is the Wigner-Seitz radius, whereas the interaction energy operator scales as $1/r_s$. Hence the kinetic energy dominates at high density (small r_s), while the interaction energy dominates at low density (large r_s).

The limit of high density is where jellium most resembles a noninteracting free electron gas. To minimize the kinetic energy, the single-electron states are delocalized plane waves, with the lowest-momentum plane-wave states being doubly occupied by spin-up and spin-down electrons, giving a paramagnetic Fermi fluid.

At lower densities, where the interaction energy is more important, it is energetically advantageous for the electron gas to spin-polarize (i.e., to have an imbalance in the number of spin-up and spin-down electrons), resulting in a ferromagnetic Fermi fluid. This phenomenon is known as *itinerant ferromagnetism*. At sufficiently low density, the kinetic-energy penalty resulting from the need to occupy higher-momentum plane-wave states is more than offset by the reduction in the interaction energy due to the fact that exchange effects keep indistinguishable electrons away from one another.

A further reduction in the interaction energy (at the expense of kinetic energy) can be achieved by localizing the electron orbitals. As a result, jellium at zero temperature at a sufficiently low density will form a so-called Wigner crystal, in which the single-particle orbitals are of approximately Gaussian form centered on crystal lattice sites. Once a Wigner crystal has formed, there may in principle be further phase transitions between different crystal structures and between different magnetic states for the Wigner crystals (e.g., antiferromagnetic to ferromagnetic spin configurations) as the density is lowered. When Wigner crystallization occurs, jellium acquires a band gap.

Within Hartree-Fock theory, the ferromagnetic fluid abruptly becomes more stable than the paramagnetic fluid at a density parameter of $r_s = 5.45$ in three dimensions (3D) and 2.01 in two dimensions (2D). However, according to Hartree-Fock theory, Wigner crystallization occurs at $r_s = 4.5$ in 3D and 1.44 in 2D, so that jellium would crystallise before itinerant ferromagnetism occurs. Furthermore, Hartree-Fock theory predicts exotic magnetic behavior, with the paramagnetic fluid being unstable to the formation of a spiral spin-density wave. Unfortunately Hartree-Fock theory does not include any description of correlation effects, which are energetically important at all but the very highest densities, and so a more accurate level of theory is required to make quantitative statements about the phase diagram of jellium.

Quantum Monte Carlo (QMC) methods, which provide an explicit treatment of electron correlation effects, are generally agreed to provide the most accurate quantitative approach for determining the zero-temperature phase diagram of jellium. The first application of the diffusion Monte Carlo method was Ceperley and Alder's famous 1980 calculation of the zero-temperature phase diagram of 3D jellium. They calculated the paramagnetic-ferromagnetic fluid transition to occur at $r_s = 75(5)$ and Wigner crystallization (to a body-centered cubic crystal) to occur at $r_s = 100(20)$. Subsequent QMC calculations have refined their phase diagram: there is a second-order transition from a paramagnetic fluid state to a partially spin-polarized fluid from $r_s = 50(2)$ to about 100 ; and Wigner crystallization occurs at $r_s = 106(1)$.

In 2D, QMC calculations indicate that the paramagnetic fluid to ferromagnetic fluid transition and Wigner crystallization occur at similar density parameters, in the range $30 < r_s < 40$. The most recent QMC calculations indicate that there is no region of stability for a ferromagnetic fluid. Instead there is a transition from a paramagnetic fluid to a hexagonal Wigner crystal at $r_s = 31(1)$. There is possibly a small region of stability for a (frustrated) antiferromagnetic Wigner crystal, before a further transition to a ferromagnetic crystal. The crystallization transition in 2D is not first order, so there must be a continuous series of transitions from fluid to crystal, perhaps involving striped crystal/fluid phases. Experimental results for a 2D hole gas in a GaAs/AlGaAs heterostructure (which, despite being clean, may not correspond exactly to the idealized jellium model) indicate a Wigner crystallization density of $r_s = 35.1(9)$.

Applications

Jellium is the simplest model of interacting electrons. It is employed in the calculation of properties of metals, where the core electrons and the nuclei are modeled as the uniform positive background and the valence electrons are treated with full rigor. Semi-infinite jellium slabs are used to investigate surface properties such as work function and surface effects such as adsorption; near surfaces the electronic density varies in an oscillatory manner, decaying to a constant value in the bulk.

Within density functional theory, jellium is used in the construction of the local-density approximation, which in turn is a component of more sophisticated exchange-correlation energy functionals. From quantum Monte Carlo calculations of jellium, accurate values of the correlation energy density have been obtained for several values of the electronic density, which have been used to construct semi-empirical correlation functionals.

The jellium model has been applied to superatoms, and used in nuclear physics.

Quantum Chemistry Composite Methods

Quantum chemistry composite methods (also referred to as thermochemical recipes) are computational chemistry methods that aim for high accuracy by combining the results of several calculations. They combine methods with a high level of theory and a small basis set with methods that employ lower levels of theory with larger basis sets. They are commonly used to calculate thermodynamic quantities such as enthalpies of formation, atomization energies, ionization energies and electron affinities. They aim for chemical accuracy which is usually defined as within 1 kcal/mol of the experimental value. The first systematic model chemistry of this type with broad applicability was called Gaussian-1 (G1) introduced by John Pople. This was quickly replaced by the Gaussian-2 (G2) which has been used extensively. The Gaussian-3 (G3) was introduced later.

Gaussian-n Theories

Gaussian-2 (G2)

The G2 uses seven calculations:

1. the molecular geometry is obtained by a MP2 optimization using the 6-31G(d) basis set and all electrons included in the perturbation. This geometry is used for all subsequent calculations.

2. The highest level of theory is a quadratic configuration interaction calculation with single and double excitations and a triples excitation contribution (QCIS-D(T)) with the 6-311G(d) basis set. Such a calculation in the Gaussian and Spartan programs also give the MP2 and MP4 energies which are also used.

3. The effect of polarization functions is assessed using an MP4 calculation with the 6-311G(2df,p) basis set.

4. The effect of diffuse functions is assessed using an MP4 calculation with the 6-311+G(d, p) basis set.

5. The largest basis set is 6-311+G(3df,2p) used at the MP2 level of theory.

6. A Hartree–Fock geometry optimization with the 6-31G(d) basis set used to give a geometry for:

7. A frequency calculation with the 6-31G(d) basis set to obtain the zero-point vibrational energy (ZPVE)

The various energy changes are assumed to be additive so the combined energy is given by:

EQCISD(T) from 2 + [EMP4 from 3 - EMP4 from 2] + [EMP4 from 4 - EMP4 from 2] + [EMP2 from 5 + EMP2 from 2 - EMP2 from 3 - EMP2 from 4]

The second term corrects for the effect of adding the polarization functions. The third term corrects for the diffuse functions. The final term corrects for the larger basis set with the terms from steps 2, 3 and 4 preventing contributions from being counted twice. Two final corrections are made to this energy. The ZPVE is scaled by 0.8929. An empirical correction is then added to account for factors not considered above. This is called the higher level correction (HC) and is given by -0.00481 x (number of valence electrons) -0.00019 x (number of unpaired valence electrons). The two numbers are obtained calibrating the results against the experimental results for a set of molecules. The scaled ZPVE and the HLC are added to give the final energy. For some molecules containing one of the third row elements Ga–Xe, a further term is added to account for spin orbit coupling.

Several variants of this procedure have been used. Removing steps 3 and 4 and relying only on the MP2 result from step 5 is significantly cheaper and only slightly less accurate. This is the G2MP2 method. Sometimes the geometry is obtained using a density functional theory method such as B3LYP and sometimes the QCISD(T) method in step 2 is replaced by the coupled cluster method CCSD(T).

The G2(+) variant, where the "+" symbol refers to added diffuse functions, better describes anions than conventional G2 theory. The 6-31+G(d) basis set is used in place of the 6-31G(d) basis set for both the initial geometry optimization, as well as the second geometry optimization and frequency calculation. Additionally, the frozen-core approximation is made for the initial MP2 optimization, whereas G2 usually uses the full calculation.

Gaussian-3 (G3)

The G3 is very similar to G2 but learns from the experience with G2 theory. The 6-311G basis set is replaced by the smaller 6-31G basis. The final MP2 calculations use a larger basis set, generally just called G3large, and correlating all the electrons not just the valence electrons as in G2 theory, additionally a spin-orbit correction term and an empirical correction for valence electrons are introduced. This gives some core correlation contributions to the final energy. The HLC takes the same form but with different empirical parameters.

Gaussian-4 (G4)

Gaussian 4 (G4) theory is an approach for the calculation of energies of molecular species containing first-row (Li–F), second-row (Na–Cl), and third row main group elements. G4 theory is an improved modification of the earlier approach G3 theory. The modifications to G3- theory are the change in an estimate of the Hartree–Fock energy limit, an expanded polarization set for the large basis set calculation, use of CCSD(T) energies, use of geometries from density functional theory and zero-point energies, and two added higher level correction parameters. According to the developers, this theory gives significant improvement over G3-theory.

Feller-Peterson-Dixon Approach (FPD)

Unlike fixed-recipe, "model chemistries", the FPD approach consists of a flexible sequence of (up to) 13 components that vary with the nature of the chemical system under study and the desired accuracy in the final results. In most instances, the primary component relies on coupled cluster theory, such as CCSD(T), or configuration interaction theory combined with large Gaussian basis sets and extrapolation to the complete basis set limit. As with some other approaches, additive corrections for core/valence, scalar relativistic and higher order correlation effects are usually included. Attention is paid to the uncertainties associated with each of the components so as to permit a crude estimate of the uncertainty in the overall results. Accurate structural parameters and vibrational frequencies are a natural byproduct of the method. While the computed molecular properties can be highly accurate, the computationally intensive nature of the FPD approach limits the size of the chemical system to which it can be applied to roughly 10 or fewer first/second row atoms.

T1

The T1 method. is an efficient computational approach developed for calculating accurate heats of formation of uncharged, closed-shell molecules comprising H, C, N, O, F, Si, P, S, Cl and Br, within experimental error. It is practical for molecules up to molecular weight ~ 500 a.m.u.

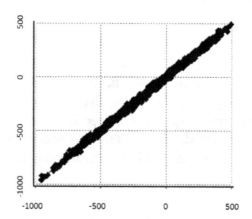

The calculated T1 heat of formation (y axis) compared to the experimental heat of formation (x axis) for a set of >1800 diverse organic molecules from the NIST thermochemical database with mean absolute and RMS errors of 8.5 and 11.5 kJ/mol, respectively.

T1 method as incorporated in Spartan consists of:

1. HF/6-31G* optimization.

2. RI-MP2/6-311+G(2d,p)[6-311G*] single point energy with dual basis set.

3. An empirical correction using atom counts, Mulliken bond orders, HF/6-31G* and RI-MP2 energies as variables.

T1 follows the G3(MP2) recipe, however, by substituting an HF/6-31G* for the MP2/6-31G* geometry, eliminating both the HF/6-31G* frequency and QCISD(T)/6-31G* energy and approximating the MP2/G3MP2large energy using dual basis set RI-MP2 techniques, the T1 method reduces computation time by up to 3 orders of magnitude. Atom counts, Mulliken bond orders and HF/6-31G* and RI-MP2 energies are introduced as variables in a linear regression fit to a set of 1126 G3(MP2) heats of formation. The T1 procedure reproduces these values with mean absolute and RMS errors of 1.8 and 2.5 kJ/mol, respectively. T1 reproduces experimental heats of formation for a set of 1805 diverse organic molecules from the NIST thermochemical database with mean absolute and RMS errors of 8.5 and 11.5 kJ/mol, respectively.

Correlation Consistent Composite Approach (ccCA)

This approach, developed at the University of North Texas by Angela K. Wilson's research group, utilizes the correlation consistent basis sets developed by Dunning and co-workers. Unlike the Gaussian-n methods, ccCA does not contain any empirically fitted term. The B3LYP density functional method with the cc-pVTZ basis set, and cc-pV(T+d)Z for third row elements (Na - Ar), are used to determine the equilibrium geometry. Single point calculations are then used to find the reference energy and additional contributions to the energy. The total ccCA energy for main group is calculated by:

$$E_{ccCA} = E_{MP2/CBS} + \Delta E_{CC} + \Delta E_{CV} + \Delta E_{SR} + \Delta E_{ZPE} + \Delta E_{SO}$$

The reference energy $E_{MP2/CBS}$ is the MP2/aug-cc-pVnZ (where n=D,T,Q) energies extrapolated at the complete basis set limit by the Peterson mixed gaussian exponential extrapolation scheme. CCSD(T)/cc-pVTZ is used to account for correlation beyond the MP2 theory:

$$\Delta E_{CC} = E_{CCSD(T)/cc\text{-}pVTZ} - E_{MP2/cc\text{-}pVTZ}$$

Core-core and core-valence interactions are accounted for using MP2(FC1)/aug-cc-pCVTZ:

$$\Delta E_{CV} = E_{MP2(FC1)/aug\text{-}cc\text{-}pCVTZ} - E_{MP2/aug\text{-}cc\text{-}pVTZ}$$

Scalar relativistic effects are also taken into account with a one-particle Douglass Kroll Hess Hamiltonian and recontracted basis sets:

$$\Delta E_{SR} = E_{MP2\text{-}DK/cc\text{-}pVTZ\text{-}DK} - E_{MP2/cc\text{-}pVTZ}$$

The last two terms are Zero Point Energy corrections scaled with a factor of 0.989 to account for deficiencies in the harmonic approximation and spin-orbit corrections considered only for atoms.

The Correlation Consistent Composite Approach is available as a keyword in NWChem and GAMESS (ccCA-S4 and ccCA-CC(2,3))

Complete Basis Set Methods (CBS)

The Complete Basis Set (CBS) methods are a family of composite methods, the members of which are: CBS-4M, CBS-QB3, and CBS-APNO, in increasing order of accuracy. These methods offer errors of 2.5, 1.1, and 0.7 kcal/mol when tested against the G2 test set. The CBS methods were developed by George Petersson and coworkers, and they make extrapolate several single-point energies to the "exact" energy. In comparison, the Gaussian-n methods perform their approximation using additive corrections. Similar to the modified G2(+) method, CBS-QB3 has been modified by the inclusion of diffuse functions in the geometry optimization step to give CBS-QB3(+). The CBS family of methods is available via keywords in the Gaussian 09 suite of programs.

Weizmann-n Theories

The Weizmann-n ab initio methods (Wn, n = 1–4) are highly accurate composite theories devoid of empirical parameters. These theories are capable of sub-kJ/mol accuracies in prediction of fundamental thermochemical quantities such as heats of formation and atomization energies, and unprecedented accuracies in prediction of spectroscopic constants. The ability of these theories to successfully reproduce the CCSD(T)/CBS (W1 and W2), CCSDT(Q)/CBS (W3), and CCSDTQ5/CBS (W4) energies relies on judicious combination of very large Gaussian basis sets with basis-set extrapolation techniques. Thus, the high accuracy of Wn theories comes with the price of a significant computational cost. In practice, for systems consisting of more than ~9 non-hydrogen atoms (with C1 symmetry), even the computationally more economical W1 theory becomes prohibitively expensive with current mainstream server hardware.

In an attempt to extend the applicability of the Wn ab initio thermochemistry methods, explicitly correlated versions of these theories have been developed: Wn-F12 (n = 1–3) and more recently even a W4-F12 theory. W1-F12 was successfully applied to large hydrocarbons (e.g., dodecahedrane, as well as to systems of biological relevance (e.g., DNA bases). W4-F12 theory has been applied to systems as large as benzene.

References

- Herzberg, Gerhard (1971). The spectra and structures of simple free radicals. New York: Dover. ISBN 0-486-65821-X.

- Harris, Daniel C.; Michael D. Bertolucci (1978). Symmetry and spectroscopy. New York: Dover. ISBN 0-486-66144-X.

- Bernath, Peter F. (1995). Spectra of Atoms and Molecules (Topics in Physical Chemistry). Oxford: Oxford University Press. ISBN 0-19-507598-6.

- Atkins, P. W.; R. S. Friedman (1999). Molecular Quantum Mechanics. Oxford: Oxford University Press. ISBN 0-19-855947-X.

- Pauli, W. (2000), Wave Mechanics: Volume 5 of Pauli Lectures on Physics (Dover Books on Physics). ISBN 978-0486414621 ; Section 44.

- Fundamentals of Molecular Symmetry by Philip R. Bunker and Per Jensen (Institute of Physics Publishing 2005) ISBN 0-7503-0941-5

- Gross, E. K. U.; Runge, E.; Heinonen, O. (1991). Many-Particle Theory. Bristol: Verlag Adam Hilger. pp. 79–80. ISBN 0-7503-0155-4.

- Giuliani, Gabriele; Vignale; Giovanni (2005). Quantum Theory of the Electron Liquid. Cambridge University Press. pp. 13–16. ISBN 978-0-521-82112-4.

- Giuliani, Gabriele; Vignale; Giovanni (2005). Quantum Theory of the Electron Liquid. Cambridge University Press. ISBN 978-0-521-82112-4.

Electronic Structure and Theories

Chemical bonding is related to the sharing or transference of electrons between atoms and in an attempt to grasp the atomic workings in the bonding of atoms, numerous models and theories have been developed. This chapter illustrates models like VSEPR theory, valence bond theory, molecular orbital theory, frontier molecular orbital theory, Ligand field theory, crystal field theory, Bohr model and old quantum theory.

VSEPR Theory

Valence shell electron pair repulsion (VSEPR) theory is a model used in chemistry to predict the geometry of individual molecules from the number of electron pairs surrounding their central atoms. It is also named the Gillespie-Nyholm theory after its two main developers. The acronym "VSEPR" is pronounced either "*ves*-per" or "vuh-*seh*-per" by some chemists.

The premise of VSEPR is that the valence electron pairs surrounding an atom tend to repel each other, and will therefore adopt an arrangement that minimizes this repulsion, thus determining the molecule's geometry. Gillespie has emphasized that the electron-electron repulsion due to the Pauli exclusion principle is more important in determining molecular geometry than the electrostatic repulsion.

VSEPR theory is based on observable electron density rather than mathematical wave functions and hence unrelated to orbital hybridisation, although both address molecular shape. While it is mainly qualitative, VSEPR has a quantitative basis in quantum chemical topology (QCT) methods such as the electron localization function and the quantum theory of atoms in molecules (QTAIM).

History

The idea of a correlation between molecular geometry and number of valence electrons (both shared and unshared) was originally proposed in 1939 by Ryutaro Tsuchida in Japan, and was independently presented in a Bakerian Lecture in 1940 by Nevil Sidgwick and Herbert Powell of the University of Oxford. In 1957, Ronald Gillespie and Ronald Sydney Nyholm of University College London refined this concept into a more detailed theory, capable of choosing between various alternative geometries.

Overview

VSEPR theory is used to predict the arrangement of electron pairs around non-hydrogen atoms in molecules, especially simple and symmetric molecules, where these key, central atoms participate in bonding to two or more other atoms; the geometry of these key atoms and their non-bonding electron pairs in turn determine the geometry of the larger whole.

The number of electron pairs in the valence shell of a central atom is determined after drawing the Lewis structure of the molecule, and expanding it to show all bonding groups and lone pairs of electrons. In VSEPR theory, a double bond or triple bond are treated as a single bonding group. The sum of the number of atoms bonded to a central atom and the number of lone pairs formed by its nonbonding valence electrons is known as the central atom's steric number.

The electron pairs (or groups if double, triple... bonds are present) are assumed to lie on the surface of a sphere centered on the central atom and tend to occupy positions that minimize their mutual repulsions by maximizing the distance between them. The number of electron pairs (or groups), therefore, determines the overall geometry that they will adopt. For example, when there are two electron pairs surrounding the central atom, their mutual repulsion is minimal when they lie at opposite poles of the sphere. Therefore, the central atom is predicted to adopt a *linear* geometry. If there are 3 electron pairs surrounding the central atom, their repulsion is minimized by placing them at the vertices of an equilateral triangle centered on the atom. Therefore, the predicted geometry is *trigonal*. Likewise, for 4 electron pairs, the optimal arrangement is *tetrahedral*.

Degree of Repulsion

The overall geometry is further refined by distinguishing between *bonding* and *nonbonding* electron pairs. The bonding electron pair shared in a sigma bond with an adjacent atom lies further from the central atom than a nonbonding (lone) pair of that atom, which is held close to its positively charged nucleus. VSEPR theory therefore views repulsion by the lone pair to be greater than the repulsion by a bonding pair. As such, when a molecule has 2 interactions with different degrees of repulsion, VSEPR theory predicts the structure where lone pairs occupy positions that allow them to experience less repulsion. Lone pair–lone pair (lp–lp) repulsions are considered stronger than lone pair–bonding pair (lp–bp) repulsions, which in turn are considered stronger than bonding pair–bonding pair (bp–bp) repulsions, distinctions that then guide decisions about overall geometry when 2 or more non-equivalent positions are possible.For instance, when 5 valence electron pairs surround a central atom, they adopt a trigonal bipyramidal molecular geometry with two collinear *axial* positions and three *equatorial* positions. An electron pair in an axial position has three close equatorial neighbors only 90° away and a fourth much farther at 180°, while an equatorial electron pair has only two adjacent pairs at 90° and two at 120°. The repulsion from the close neighbors at 90° is more important, so that the axial positions experience more repulsion than the equatorial positions; hence, when there are lone pairs, they tend to occupy equatorial positions as shown in the diagrams of the next section for steric number five.

The difference between lone pairs and bonding pairs may also be used to rationalize deviations from idealized geometries. For example, the H_2O molecule has four electron pairs in its valence shell: two lone pairs and two bond pairs. The four electron pairs are spread so as to point roughly towards the apices of a tetrahedron. However, the bond angle between the two O–H bonds is only 104.5°, rather than the 109.5° of a regular tetrahedron, because the two lone pairs (whose density or probability envelopes lie closer to the oxygen nucleus) exert a greater mutual repulsion than the two bond pairs.

An advanced-level explanation replaces the above distinction with two rules:

- Bent's rule: An electron pair of a more electropositive ligand constitutes greater repulsion. This explains why the Cl in $PClF_4$ prefers the equatorial position and why the bond angle in oxygen difluoride (103.8°) is smaller than that of water (104.5°). Lone pairs are then considered to be a special case of this rule, held by a "ghost ligand" in the limit of electropositivity.

- A higher bond order constitutes greater repulsion. This explains why in phosgene, the oxygen–chlorine bond angle (124.1°) is larger than the chlorine–chlorine bond angle (111.8°) even though chlorine is more electropositive than oxygen. In the carbonate ion, all three bond angles are equivalent due to resonance.

AXE Method

The "AXE method" of electron counting is commonly used when applying the VSEPR theory. The A represents the central atom and always has an implied subscript one. The X represents each of ligands (atoms bonded to A). The E represents the number of lone electron *pairs* surrounding the central atom. The sum of X and E is known as the steric number.

Based on the steric number and distribution of Xs and Es, VSEPR theory makes the predictions in the following tables. Note that the geometries are named according to the atomic positions only and not the electron arrangement. For example, the description of AX_2E_1 as a bent molecule means that the three atoms AX_2 are not in one straight line, although the lone pair helps to determine the geometry.

Steric-number	Molecular geometry 0 lone pairs	Molecular geometry 1 lone pair	Molecular geometry 2 lone pairs	Molecular geometry 3 lone pairs
2	X—A—X Linear (CO_2)			

3	Trigonal planar (BCl_3)	Bent (SO_2)		
4	Tetrahedral (CH_4)	Trigonal pyramidal (NH_3)	Bent (H_2O)	
5	Trigonal bipyramidal (PCl_5)	Seesaw (SF_4)	T-shaped (ClF_3)	Linear (I-3)
6	Octahedral (SF_6)	Square pyramidal (BrF_5)	Square planar (XeF_4)	
7	Pentagonal bipyramidal (IF_7)	Pentagonal pyramidal (XeOF–5)	Pentagonal planar (XeF–5)	

8	Square antiprismatic (TaF3−8)			
9	Tricapped trigonal prismatic (ReH2−9) *or* Capped square antiprismatic			
10	Bicapped square anti-prismatic *or* Bicapped dodecadeltahedral			
11	Octadecahedral			
12	Icosahedral			
14	Bicapped hexagonal antiprismatic			

Molecule type	Shape	Electron arrangement including lone pairs, shown in pale yellow	Geometry excluding lone pairs	Examples
AX_2E_0	Linear			$BeCl_2$, $HgCl_2$, CO_2
AX_2E_1	Bent			$NO-2$, SO_2 O_3, CCl_2
AX_2E_2	Bent			H_2O, OF_2
AX_2E_3	Linear			XeF_2, $I-3$, $XeCl_2$
AX_3E_0	Trigonal planar			BF_3, $CO2-3$, $NO-3$, SO_3
AX_3E_1	Trigonal pyramidal			NH_3, PCl_3

AX_3E_2	T-shaped			ClF_3, BrF_3
AX_4E_0	Tetrahedral			CH_4, $PO3-4$, $SO2-4$, $ClO-4$, XeO_4
AX_4E_1	Seesaw or disphenoidal			SF_4
AX_4E_2	Square planar			XeF_4
AX_5E_0	Trigonal bipyramidal			PCl_5
AX_5E_1	Square pyramidal			ClF_5, BrF_5, $XeOF_4$
AX_5E_2	Pentagonal planar			$XeF-5$
AX_6E_0	Octahedral			SF_6, WCl_6

AX$_6$E$_1$	Pentagonal pyrami-dal			XeOF-5, IOF2-5
AX$_7$E$_0$	Pentagonal bipyra-midal			IF$_7$
AX$_8$E$_0$	Square antiprismat-ic			IF-8, ZrF4-8, ReF-8
AX$_9$E$_0$	Tricapped trigo-nal prismatic (as drawn) or capped square antiprismat-ic			ReH2-9

When the substituent (X) atoms are not all the same, the geometry is still approximately valid, but the bond angles may be slightly different from the ones where all the outside atoms are the same. For example, the double-bond carbons in alkenes like C_2H_4 are AX$_3$E$_0$, but the bond angles are not all exactly 120°. Likewise, $SOCl_2$ is AX$_3$E$_1$, but because the X substituents are not identical, the X–A–X angles are not all equal.

As a tool in predicting the geometry adopted with a given number of electron pairs, an often used physical demonstration of the principle of minimal electron pair repulsion utilizes inflated balloons. Through handling, balloons acquire a slight surface electrostatic charge that results in the adoption of roughly the same geometries when they are tied together at their stems as the corresponding number of electron pairs. For example, five balloons tied together adopt the trigonal bipyramidal geometry, just as do the five bonding pairs of a PCl_5 molecule (AX$_5$) or the two bonding and three non-bonding pairs of a XeF_2 molecule (AX$_2$E$_3$). The molecular geometry of the former is also trigonal bipyramidal, whereas that of the latter is linear.

Examples

The methane molecule (CH_4) is tetrahedral because there are four pairs of electrons.

The four hydrogen atoms are positioned at the vertices of a tetrahedron, and the bond angle is $\cos^{-1}(-\frac{1}{3}) \approx 109°\ 28'$. This is referred to as an AX_4 type of molecule. As mentioned above, A represents the central atom and X represents an outer atom.

The ammonia molecule (NH_3) has three pairs of electrons involved in bonding, but there is a lone pair of electrons on the nitrogen atom. It is not bonded with another atom; however, it influences the overall shape through repulsions. As in methane above, there are four regions of electron density. Therefore, the overall orientation of the regions of electron density is tetrahedral. On the other hand, there are only three outer atoms. This is referred to as an AX_3E type molecule because the lone pair is represented by an E. By definition, the molecular shape or geometry describes the geometric arrangement of the atomic nuclei only, which is trigonal-pyramidal for NH_3.

Steric numbers of 7 or greater are possible, but are less common. The steric number of 7 occurs in iodine heptafluoride (IF_7); the base geometry for a steric number of 7 is pentagonal bipyramidal. The most common geometry for a steric number of 8 is a square antiprismatic geometry. Examples of this include the octacyanomolybdate ($Mo(CN)4-8$) and octafluorozirconate ($ZrF4-8$) anions.

The nonahydridorhenate ion ($ReH2-9$) in potassium nonahydridorhenate is a rare example of a compound with a steric number of 9, which has a tricapped trigonal prismatic geometry. Another example is the octafluoroxenate ion ($XeF2-8$) in nitrosonium octafluoroxenate(VI), although in this case one of the electron pairs is a lone pair, and therefore the molecule actually has a distorted square antiprismatic geometry.

Possible geometries for steric numbers of 10, 11, 12, or 14 are bicapped square antiprismatic (or bicapped dodecadeltahedral), octadecahedral, icosahedral, and bicapped hexagonal antiprismatic, respectively. No compounds with steric numbers this high involving monodentate ligands exist, and those involving multidentate ligands can often be analysed more simply as complexes with lower steric numbers when some multidentate ligands are treated as a unit.

Exceptions

There are groups of compounds where VSEPR fails to predict the correct geometry.

Some AX_2E_0 Molecules

The gas phase structures of the triatomic halides of the heavier members of group 2, (i.e., calcium, strontium and barium halides, MX_2), are not linear as predicted but are bent, (approximate X–M–X angles: CaF_2, 145°; SrF_2, 120°; BaF_2, 108°; $SrCl_2$, 130°; $BaCl_2$, 115°; $BaBr_2$, 115°; BaI_2, 105°). It has been proposed by Gillespie that this is caused by interaction of the ligands with the electron core of the metal atom, polarising it so that the inner shell is not spherically symmetric, thus influencing the molecular geometry.

Ab initio calculations have been cited to propose that contributions from d orbitals in the shell below the valence shell are responsible. Disilynes are also bent, despite having no lone pairs.

Some AX_2E_2 Molecules

One example of the AX_2E_2 geometry is molecular lithium oxide, Li_2O, a linear rather than bent structure, which is ascribed to its bonds being essentially ionic and the strong lithium-lithium repulsion that results. Another example is $O(SiH_3)_2$ with an Si–O–Si angle of 144.1°, which compares to the angles in Cl_2O (110.9°), $(CH_3)_2O$ (111.7°), and $N(CH_3)_3$ (110.9°). Gillespie and Robinson rationalize the Si–O–Si bond angle based on the observed ability of a ligand's lone pair to most greatly repel other electron pairs when the ligand electronegativity is greater than or equal to that of the central atom. In $O(SiH_3)_2$, the central atom is more electronegative, and the lone pairs are less localized and more weakly repulsive. The larger Si–O–Si bond angle results from this and strong ligand-ligand repulsion by the relatively large -SiH_3 ligand.

Some AX_6E_1 and AX_8E_1 Molecules

Some AX_6E_1 molecules, e.g. xenon hexafluoride (XeF_6) and the Te(IV) and Bi(III) anions, $TeCl_6^{2-}$, $TeBr_6^{2-}$, $BiCl_6^{3-}$, $BiBr_6^{3-}$ and BiI_6^{3-}, are octahedra, rather than pentagonal pyramids, and the lone pair does not affect the geometry to the degree predicted by VSEPR. One rationalization is that steric crowding of the ligands allows little or no room for the non-bonding lone pair; another rationalization is the inert pair effect.

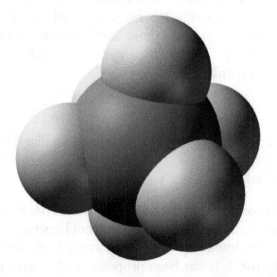

Xenon hexafluoride, which has a distorted octahedral geometry.

Transition Metal Molecules

Many transition metal compounds have unusual geometries, which can be ascribed to ligand bonding interaction with the d subshell and to absence of valence shell lone pairs. Gillespie suggested that this interaction can be weak or strong. Weak interaction is dealt with by the Kepert model, while strong interaction produces bonding pairs that also occupy the respective antipodal points of the sphere. This is similar to predictions based on sd hybrid orbitals using the VALBOND theory. The repulsion of these bidirectional bonding pairs leads to a different prediction of shapes.

Hexamethyltungsten, a transition metal compound whose geometry is different from main group coordination.

Molecule type	Shape	Geometry	Examples
AX_2	Bent		$VO+2$
AX_3	Trigonal pyramidal		CrO_3
AX_4	Tetrahedral		$TiCl_4$
AX_5	Square pyramidal		$Ta(CH_3)_5$

AX$_6$	Trigonal prismatic	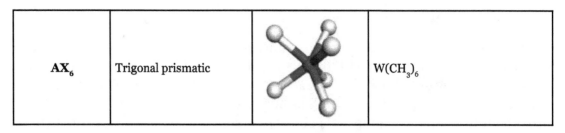	W(CH$_3$)$_6$

The square planar shape associated with a d^8 electronic configuration is an exception to the Kepert model. This can be rationalized by considering the increased crystal field stabilization energy as compared to a tetrahedral geometry.

Odd-electron Molecules

The VSEPR theory can be extended to molecules with an odd number of electrons by treating the unpaired electron as a "half electron pair" — for example, Gillespie and Nyholm suggested that the decrease in the bond angle in the series NO$_2^+$(180°), NO$_2$ (134°), NO$_2^-$ (115°) indicates that a given set of bonding electron pairs exert a weaker repulsion on a single non-bonding electron than on a pair of non-bonding electrons. In effect, they considered nitrogen dioxide as an AX$_2$E$_{0.5}$ molecule with a geometry intermediate between NO$_2^+$ and NO$_2^-$. Similarly chlorine dioxide (ClO$_2$) is AX$_2$E$_{1.5}$ with a geometry intermediate between ClO$_2^+$ and ClO$_2^-$.

Finally the methyl radical (CH$_3$) is predicted to be trigonal pyramidal like the methyl anion (CH$_3^-$), but with a larger bond angle as in the trigonal planar methyl cation (CH$_3^+$). However, in this case the VSEPR prediction is not quite true, as CH$_3$ is actually planar, although its distortion to a pyramidal geometry requires very little energy.

Valence Bond Theory

In chemistry, valence bond (VB) theory is one of two basic theories, along with molecular orbital (MO) theory, that were developed to use the methods of quantum mechanics to explain chemical bonding. It focuses on how the atomic orbitals of the dissociated atoms combine to give individual chemical bonds when a molecule is formed. In contrast, molecular orbital theory has orbitals that cover the whole molecule.

History

In 1916, G. N. Lewis proposed that a chemical bond forms by the interaction of two shared bonding electrons, with the representation of molecules as Lewis structures. In 1927 the Heitler–London theory was formulated which for the first time enabled the calculation of bonding properties of the hydrogen molecule H$_2$ based on quantum mechanical considerations. Specifically, Walter Heitler determined how to use Schröding-er's wave equation (1926) to show how two hydrogen atom wavefunctions join togeth-er, with plus, minus, and exchange terms, to form a covalent bond. He then called up his associate Fritz London and they worked out the details of the theory over the course of the night. Later, Linus Pauling used the pair bonding ideas of Lewis together with Heitler–London theory to develop two other key concepts in VB theory: resonance (1928) and orbital hybridization (1930). According to Charles Coulson, author of the noted 1952 book *Valence*, this period marks the start of "modern valence bond theory", as contrasted with older valence bond theories, which are essentially electronic theo-ries of valence couched in pre-wave-mechanical terms. Resonance theory was criticized as imperfect by Soviet chemists during the 1950s.

Theory

According to this theory a covalent bond is formed between the two atoms by the overlap of half filled valence atomic orbitals of each atom containing one unpaired electron. A valence bond structure is similar to a Lewis structure, but where a single Lewis structure cannot be written, several valence bond structures are used. Each of these VB structures represents a specific Lewis structure. This combination of valence bond structures is the main point of resonance theory. Valence bond the-ory considers that the overlapping atomic orbitals of the participating atoms form a chemical bond. Because of the overlapping, it is most probable that electrons should be in the bond region. Valence bond theory views bonds as weakly cou-pled orbitals (small overlap). Valence bond theory is typically easier to employ in ground state molecules. The inner-shell orbitals and electrons remain essentially unchanged during the formation of bonds.

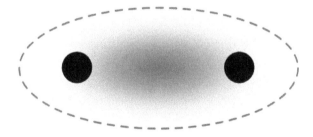

σ bond between two atoms: localization of electron density

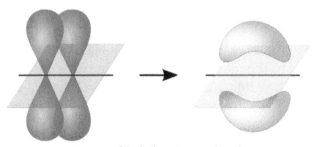

Two p-orbitals forming a π-bond.

The overlapping atomic orbitals can differ. The two types of overlapping orbitals are sigma and pi. Sigma bonds occur when the orbitals of two shared electrons overlap head-to-head. Pi bonds occur when two orbitals overlap when they are parallel. For example, a bond between two s-orbital electrons is a sigma bond, because two spheres are always coaxial. In terms of bond order, single bonds have one sigma bond, double bonds consist of one sigma bond and one pi bond, and triple bonds contain one sigma bond and two pi bonds. However, the atomic orbitals for bonding may be hybrids. Often, the bonding atomic orbitals have a character of several possible types of orbitals. The methods to get an atomic orbital with the proper character for the bonding is called hybridization.

Valence Bond Theory Today

Modern valence bond theory now complements molecular orbital theory, which does not adhere to the valence bond idea that electron pairs are localized between two specific atoms in a molecule but that they are distributed in sets of molecular orbitals which can extend over the entire molecule. Molecular orbital theory can predict magnetic and ionization properties in a straightforward manner, while valence bond theory gives similar results but is more complicated. Modern valence bond theory views aromatic properties of molecules as due to spin coupling of the π orbitals. This is essentially still the old idea of resonance between Kekulé and Dewar structures. In contrast, molecular orbital theory views aromaticity as delocalization of the π-electrons. Valence bond treatments are restricted to relatively small molecules, largely due to the lack of orthogonality between valence bond orbitals and between valence bond structures, while molecular orbitals are orthogonal. On the other hand, valence bond theory provides a much more accurate picture of the reorganization of electronic charge that takes place when bonds are broken and formed during the course of a chemical reaction. In particular, valence bond theory correctly predicts the dissociation of homonuclear diatomic molecules into separate atoms, while simple molecular orbital theory predicts dissociation into a mixture of atoms and ions. For example the molecular orbital function for dihydrogen is an equal mixture of the covalent and ionic valence bond structures and so predicts incorrectly that the molecule would dissociate into an equal mixture of hydrogen atoms and hydrogen positive and negative ions.

Modern valence bond theory replaces the overlapping atomic orbitals by overlapping valence bond orbitals that are expanded over a large number of basis functions, either centered each on one atom to give a classical valence bond picture, or centered on all atoms in the molecule. The resulting energies are more competitive with energies from calculations where electron correlation is introduced based on a Hartree–Fock reference wavefunction. The most recent text is by Shaik and Hiberty.

Applications of Valence Bond Theory

An important aspect of the VB theory is the condition of maximum overlap, which leads to the formation of the strongest possible bonds. This theory is used to explain the covalent bond formation in many molecules.

For example, in the case of the F_2 molecule, the F–F bond is formed by the overlap of p_z orbitals of the two F atoms, each containing an unpaired electron. Since the nature of the overlapping orbitals are different in H_2 and F_2 molecules, the bond strength and bond lengths differ between H_2 and F_2 molecules.

In an HF molecule the covalent bond is formed by the overlap of the $1s$ orbital of H and the $2p_z$ orbital of F, each containing an unpaired electron. Mutual sharing of electrons between H and F results in a covalent bond in HF.

Molecular Orbital Theory

In chemistry, molecular orbital (MO) theory is a method for determining molecular structure in which electrons are not assigned to individual bonds between atoms, but are treated as moving under the influence of the nuclei in the whole molecule. The spatial and energetic properties of electrons within atoms are fixed by quantum mechanics to form orbitals that contain these electrons. While atomic orbitals contain electrons ascribed to a single atom, molecular orbitals, which surround a number of atoms in a molecule, contain valence electrons between atoms. Molecular orbital theory, which was proposed in the early twentieth century, revolutionized the study of bonding by approximating the positions of bonded electrons—the molecular orbitals—as linear combinations of atomic orbitals (LCAO). These approximations are now made by applying the density functional theory (DFT) or Hartree–Fock (HF) models to the Schrödinger equation.

Quantitative Applications

In this theory, each molecule has a set of molecular orbitals, in which it is assumed that the molecular orbital wave function ψ_j can be written as a simple weighted sum of the n constituent atomic orbitals χ_i, according to the following equation:

$$\psi_j = \sum_{i=1}^{n} c_{ij} \chi_i.$$

One may determine c_{ij} coefficients numerically by substituting this equation into the Schrödinger equation and applying the variational principle. The variational principle is a mathematical technique used in quantum mechanics to build up the coefficients of each atomic orbital basis. A larger coefficient means that the orbital basis is composed more of that particular contributing atomic orbital—hence, the molecular orbital is best characterized by that type. This method of quantifying orbital contribution as Linear Combinations of Atomic Orbitals is used in computational chemistry. An additional unitary transformation can be applied on the system to accelerate the convergence in some computational schemes. Molecular orbital theory was seen as a competitor to valence bond theory in the 1930s, before it was realized that the two methods are closely related and that when extended they become equivalent.

History

Molecular orbital theory was developed, in the years after valence bond theory had been established (1927), primarily through the efforts of Friedrich Hund, Robert Mulliken, John C. Slater, and John Lennard-Jones. MO theory was originally called the Hund-Mulliken theory. According to German physicist and physical chemist Erich Hückel, the first quantitative use of molecular orbital theory was the 1929 paper of Lennard-Jones. This paper notably predicted a triplet ground state for the dioxygen molecule which explained its paramagnetism before valence bond theory, which came up with its own explanation in 1931. The word *orbital* was introduced by Mulliken in 1932. By 1933, the molecular orbital theory had been accepted as a valid and useful theory.

Erich Hückel applied molecular orbital theory to unsaturated hydrocarbon molecules starting in 1931 with his Hückel molecular orbital (HMO) method for the determination of MO energies for pi electrons, which he applied to conjugated and aromatic hydrocarbons. This method provided an explanation of the stability of molecules with six pi-electrons such as benzene.

The first accurate calculation of a molecular orbital wavefunction was that made by Charles Coulson in 1938 on the hydrogen molecule. By 1950, molecular orbitals were completely defined as eigenfunctions (wave functions) of the self-consistent field Hamiltonian and it was at this point that molecular orbital theory became fully rigorous and consistent. This rigorous approach is known as the Hartree–Fock method for molecules although it had its origins in calculations on atoms. In calculations on molecules, the molecular orbitals are expanded in terms of an atomic orbital basis set, leading to the Roothaan equations. This led to the development of many ab initio quantum chemistry methods. In parallel, molecular orbital theory was applied in a more approximate manner using some empirically derived parameters in methods now known as semi-empirical quantum chemistry methods.

The success of Molecular Orbital Theory also spawned ligand field theory, which was developed during the 1930s and 1940s as an alternative to crystal field theory.

Types of Orbitals

Molecular orbital (MO) theory uses a linear combination of atomic orbitals (LCAO) to represent molecular orbitals resulting from bonds between atoms. These are often divided into *bonding* orbitals, anti-bonding orbitals, and non-bonding orbitals. A bonding orbital concentrates electron density in the region *between* a given pair of atoms, so that its electron density will tend to attract each of the two nuclei toward the other and hold the two atoms together. An anti-bonding orbital concentrates electron density "behind" each nucleus (i.e. on the side of each atom which is farthest from the other atom), and so tends to pull each of the two nuclei away from the other and actually weaken the bond between the two nuclei. Electrons in non-bonding orbitals tend to be associated with atomic orbitals that do not interact positively or negatively with one another, and electrons in these orbitals neither contribute to nor detract from bond strength.

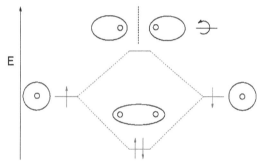

MO diagram showing the formation of molecular orbitals of H_2 (centre) from atomic orbitals of two H atoms. The lower-energy MO is bonding with electron density concentrated between the two H nuclei. The higher-energy MO is anti-bonding with electron density concentrated behind each H nucleus.

Molecular orbitals are further divided according to the types of atomic orbitals they are formed from. Chemical substances will form bonding interactions if their orbitals become lower in energy when they interact with each other. Different bonding orbitals are distinguished that differ by electron configuration (electron cloud shape) and by energy levels.

The molecular orbitals of a molecule can be illustrated in molecular orbital diagrams.

Overview

MO theory provides a global, delocalized perspective on chemical bonding. In MO theory, *any* electron in a molecule may be found *anywhere* in the molecule, since quantum conditions allow electrons to travel under the influence of an arbitrarily large number of nuclei, as long as they are in eigenstates permitted by certain quantum rules. Thus, when excited with the requisite amount of energy through high-frequency light or other means, electrons can transition to higher-energy molecular orbitals. For instance, in the simple case of a hydrogen diatomic molecule, promotion of a single electron from a bonding orbital to an antibonding orbital can occur under UV radiation. This promotion weakens the bond between the two hydrogen atoms and can lead to photodissociation—the breaking of a chemical bond due to the absorption of light.

Although in MO theory *some* molecular orbitals may hold electrons that are more localized between specific pairs of molecular atoms, *other* orbitals may hold electrons that are spread more uniformly over the molecule. Thus, overall, bonding is far more delocalized in MO theory, which makes it more applicable to resonant molecules that have equivalent non-integer bond orders than valence bond (VB) theory. This makes MO theory more useful for the description of extended systems.

An example is the MO description of benzene, C_6H_6, which is an aromatic hexagonal ring of six carbon atoms and three double bonds. In this molecule, 24 of the 30 total valence bonding electrons—24 coming from carbon atoms and 6 coming from hydrogen atoms—are located in 12 σ (sigma) bonding orbitals, which are located mostly between pairs of atoms (C-C or C-H), similarly to the electrons in the valence bond description. However, in benzene the remaining six bonding electrons are located in three π (pi) molecular bonding orbitals that are delocalized around the ring. Two of these electrons are in an MO that has equal orbital contributions from all six atoms. The other four electrons are in orbitals with vertical nodes at right angles to each other. As in the VB theory, all of these six delocalized π electrons reside in a larger space that exists above and below the ring plane. All carbon-carbon bonds in benzene are chemically equivalent. In MO theory this is a direct consequence of the fact that the three molecular π orbitals combine and evenly spread the extra six electrons over six carbon atoms.

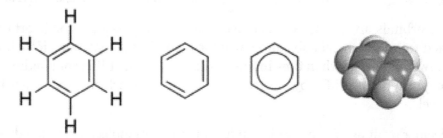

Structure of benzene

In molecules such as methane, CH_4, the eight valence electrons are found in four MOs that are spread out over all five atoms. However, it is possible to transform the MOs into four localized sp³ orbitals. Linus Pauling, in 1931, hybridized the carbon 2s and 2p orbitals so that they pointed directly at the hydrogen 1s basis functions and featured maximal overlap. However, the delocalized MO description is more appropriate for predicting ionization energies and the positions of spectral absorption bands. When methane is ionized, a single electron is taken from the valence MOs, which can come from the s bonding or the triply degenerate p bonding levels, yielding two ionization energies. In comparison, the explanation in VB theory is more complicated. When one electron is removed from an sp³ orbital, resonance is invoked between four valence bond structures, each of which has a single one-electron bond and three two-electron bonds. Triply degenerate T_2 and A_1 ionized states (CH_4^+) are produced from different linear combinations of these four structures. The difference in energy between the ionized and ground state gives the two ionization energies.

As in benzene, in substances such as beta carotene, chlorophyll, or heme, some electrons in the π orbitals are spread out in molecular orbitals over long distances in a molecule, resulting in light absorption in lower energies (the visible spectrum), which accounts for the characteristic colours of these substances. This and other spectroscopic data for molecules are well explained in MO theory, with an emphasis on electronic states associated with multicenter orbitals, including mixing of orbitals premised on principles of orbital symmetry matching. The same MO principles also naturally explain some electrical phenomena, such as high electrical conductivity in the planar direction of the hexagonal atomic sheets that exist in graphite. This results from continuous band overlap of half-filled p orbitals and explains electrical conduction. MO theory recognizes that some electrons in the graphite atomic sheets are completely delocalized over arbitrary distances, and reside in very large molecular orbitals that cover an entire graphite sheet, and some electrons are thus as free to move and therefore conduct electricity in the sheet plane, as if they resided in a metal.

Frontier Molecular Orbital Theory

In chemistry, frontier molecular orbital theory is an application of MO theory describing HOMO / LUMO interactions.

History

In 1952, Kenichi Fukui published a paper in the *Journal of Chemical Physics* titled "A molecular theory of reactivity in aromatic hydrocarbons." Though widely criticized at the time, he later shared the Nobel Prize in Chemistry with Roald Hoffmann for his work on reaction mechanisms. Hoffman's work focused on creating a set of four pericyclic reactions in organic chemistry, based on orbital symmetry, which he coauthored with Robert Burns Woodward, entitled "The Conservation of Orbital Symmetry."

Fukui's own work looked at the frontier orbitals, and in particular the effects of the Highest Occupied Molecular Orbital (HOMO) and the Lowest Unoccupied Molecular Orbital (LUMO) on reaction mechanisms, which led to it being called Frontier Molecular Orbital Theory (FMO Theory). He used these interactions to better understand the conclusions of the Woodward–Hoffmann rules.

Theory

Fukui realized that a good approximation for reactivity could be found by looking at the frontier orbitals (HOMO/LUMO). This was based on three main observations of molecular orbital theory as two molecules interact:

1. The occupied orbitals of different molecules repel each other.

2. Positive charges of one molecule attract the negative charges of the other.

3. The occupied orbitals of one molecule and the unoccupied orbitals of the other (especially the HOMO and LUMO) interact with each other causing attraction.

From these observations, frontier molecular orbital (FMO) theory simplifies reactivity to interactions between the HOMO of one species and the LUMO of the other. This helps to explain the predictions of the Woodward–Hoffmann rules for thermal pericyclic reactions, which are summarized in the following statement:

"A ground-state pericyclic change is symmetry-allowed when the total number of $(4q+2)_s$ and $(4r)_a$ components is odd"

$(4q+2)_s$ refers to the number of aromatic, suprafacial electron systems; likewise, $(4r)_a$ refers to antiaromatic, antarafacial systems. It can be shown that if the total number of these systems is odd then the reaction is thermally allowed.

Applications

Cycloadditions

A cycloaddition is a reaction that simultaneously forms at least two new bonds, and in doing so, converts two or more open-chain molecules into rings. The transition states for these reactions typically involves the electrons of the molecules moving in continuous rings, making it a pericyclic reaction. These reactions can be predicted by the Woodward–Hoffmann rules and thus are closely approximated by FMO Theory.

The Diels–Alder reaction between maleic anhydride and cyclopentadiene is allowed by the Woodward–Hoffmann rules because there are six electrons moving suprafacially and no electrons moving antarafacially. Thus, there is one $(4q + 2)_s$ component and no $(4r)_a$ component, which means the reaction is allowed thermally.

FMO theory also finds that this reaction is allowed and goes even further by predicting its stereoselectivity, which is unknown under the Woodward-Hoffmann rules. Since this is a [4 + 2], the reaction can be simplified by considering the reaction between butadiene and ethene. The HOMO of butadiene and the LUMO of ethene are both antisymmetric (rotationally symmetric), meaning the reaction is allowed.

LUMO —

LUMO   LUMO

HOMO   HOMO

In terms of the stereoselectivity of the reaction between maleic anhydride and cyclopentadiene, the *endo*-product is favored, a result best explained through FMO theory. The maleic anhydride is an electron-withdrawing species that makes the dieneophile electron deficient, forcing the regular Diels–Alder reaction. Thus, only the reaction between the HOMO of cyclopentadiene and the LUMO of maleic anhydride is allowed. Furthermore, though the *exo*-product is the more thermodynamically stable isomer, there are secondary (non-bonding) orbital interactions in the *endo*- transition state, lowering its energy and making the reaction towards the *endo*- product faster, and therefore more kinetically favorable. Since the *exo*-product has primary (bonding) orbital interactions it can still form, but since the *endo*-product forms faster it is the major product.

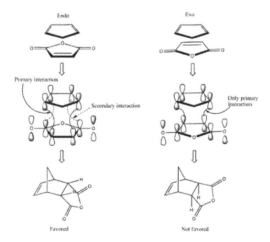

Note: The HOMO of ethene and the LUMO of butadiene are both symmetric, meaning the reaction between these species is allowed as well. This is referred to as the "inverse electron demand Diels–Alder."

Sigmatropic Reactions

A sigmatropic rearrangement is a reaction in which a sigma bond moves across a conjugated pi system with a concomitant shift in the pi bonds. The shift in the sigma bond may be antarafacial or suprafacial. In the example of a [1,5] shift in pentadiene, if there is a suprafacial shift, there is 6 e⁻ moving suprafacially and none moving antarafacially, implying this reaction is allowed by the Woodward–Hoffmann rules. For an antarafacial shift, the reaction is not allowed.

These results can be predicted with FMO theory by observing the interaction between the HOMO and LUMO of the species. To use FMO theory, the reaction should be considered as two separate ideas: (1) whether or not the reaction is allowed, and (2) which mechanism the reaction proceeds though. In the case of a [1,5] shift on pentadiene, the HOMO of the sigma bond (i.e. a constructive bond) and the LUMO of butadiene on the remaining 4 carbons is observed. Assuming the reaction happens suprafacially, the shift results with the HOMO of butadiene on the 4 carbons that are not involved in the sigma bond of the product. Since the pi system changed from the LUMO to the HOMO, this reaction is allowed (though it would not be allowed if the pi system went from LUMO to LUMO).

To explain why the reaction happens suprafacially, first notice that the terminal orbitals are in the same phase. For there to be a constructive sigma bond formed after the shift, the reaction would have to be suprafacial. If the species shifted antarafacially then it would form an antibonding orbital and there would not be a constructive sigma shift.

It is worth noting that in propene the shift would have to be antarafacial, but since the molecule is very small that twist is not possible and the reaction is not allowed.

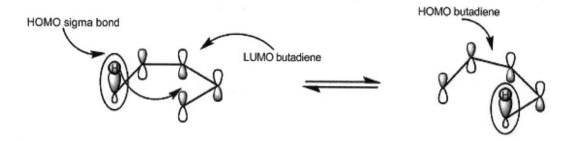

Electrocyclic Reactions

An electrocyclic reaction is a pericyclic reaction involving the net loss of a pi bond and creation of a sigma bond with formation of a ring. This reaction proceeds through either a conrotatory or disrotatory mechanism. In the conrotatory ring opening of cyclobutene, there are two electrons moving suprafacially (on the pi bond) and two moving antarafacially (on the sigma bond). This means there is one $4q + 2$ suprafacial system and no 4r antarafacial system; thus the conrotatory process is thermally allowed by the Woodward–Hoffmann rules.

The HOMO of the sigma bond (i.e. a constructive bond) and the LUMO of the pi bond are important in the FMO theory consideration. If the ring opening uses a conrotatory process then the reaction results with the HOMO of butadiene. As in the previous examples the pi system moves from a LUMO species to a HOMO species, meaning this reaction is allowed.

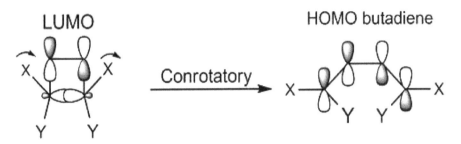

Ligand Field Theory

Ligand field theory (LFT) describes the bonding, orbital arrangement, and other characteristics of coordination complexes. It represents an application of molecular orbital theory to transition metal complexes. A transition metal ion has nine valence atomic orbitals - consisting of five nd, three $(n+1)p$, and one $(n+1)s$ orbitals. These orbitals are of appropriate energy to form bonding interaction with ligands. The LFT analysis is highly dependent on the geometry of the complex, but most explanations begin by describing octahedral complexes, where six ligands coordinate to the metal. Other complexes can be described by reference to crystal field theory.

History

Ligand field theory resulted from combining the principles laid out in molecular orbital theory and crystal field theory, which describes the loss of degeneracy of metal d orbitals in transition metal complexes. Griffith and Orgel championed ligand field theory as a more accurate description of such complexes. They used the electrostatic principles established in crystal field theory to describe transition metal ions in solution, and they used molecular orbital theory to explain the differences in metal-ligand interactions. In their paper, they propose that the chief cause of color differences in transition metal complexes in solution is the incomplete d orbital subshells. That is, the unoccupied d orbitals of transition metals participate in bonding, which influences the colors they absorb in solution. In ligand field theory, the various d orbitals are affected differently when surrounded by a field of neighboring ligands and are raised or lowered in energy based on the strength of their interaction with the ligands.

Bonding

σ-Bonding (Sigma Bonding)

In an octahedral complex, the molecular orbitals created by coordination can be seen as resulting from the donation of two electrons by each of six σ-donor ligands to the d-orbitals on the metal. In octahedral complexes, ligands approach along the x-, y- and z-axes, so their σ-symmetry orbitals form bonding and anti-bonding combinations with the d_z^2 and $d_{x^2-y^2}$ orbitals. The d_{xy}, d_{xz} and d_{yz} orbitals remain non-bonding orbitals. Some weak bonding (and anti-bonding) interactions with the s and p orbitals of the metal also occur, to make a total of 6 bonding (and 6 anti-bonding) molecular orbitals

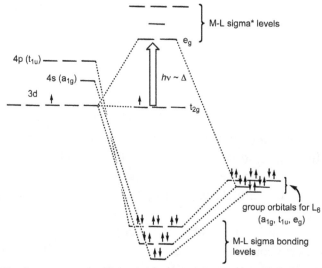

Ligand-Field scheme summarizing σ-bonding in the octahedral complex $[Ti(H_2O)_6]^{3+}$.

In molecular symmetry terms, the six lone-pair orbitals from the ligands (one from each ligand) form six symmetry adapted linear combinations (SALCs) of orbitals, also sometimes called ligand group orbitals (LGOs). The irreducible representations that these span are a_{1g}, t_{1u} and e_g. The metal also has six valence orbitals that span these irreducible representations - the s orbital is labeled a_{1g}, a set of three p-orbitals is labeled t_{1u}, and the d_z^2 and $d_{x^2-y^2}$ orbitals are labeled e_g. The six σ-bonding molecular orbitals result from the combinations of ligand SALC's with metal orbitals of the same symmetry.

π-bonding (Pi Bonding)

π bonding in octahedral complexes occurs in two ways: via any ligand p-orbitals that are not being used in σ bonding, and via any π or π* molecular orbitals present on the ligand.

In the usual analysis, the p-orbitals of the metal are used for σ bonding (and have the wrong symmetry to overlap with the ligand p or π or π* orbitals anyway), so the π interactions take place with the appropriate metal d-orbitals, i.e. d_{xy}, d_{xz} and d_{yz}. These are the orbitals that are non-bonding when only σ bonding takes place.

One important π bonding in coordination complexes is metal-to-ligand π bonding, also called π backbonding. It occurs when the LUMOs (lowest unoccupied molecular orbitals) of the ligand are anti-bonding π^* orbitals. These orbitals are close in energy to the d_{xy}, d_{xz} and d_{yz} orbitals, with which they combine to form bonding orbitals (i.e. orbitals of lower energy than the aforementioned set of d-orbitals). The corresponding anti-bonding orbitals are higher in energy than the anti-bonding orbitals from σ bonding so, after the new π bonding orbitals are filled with electrons from the metal d-orbitals, Δ_O has increased and the bond between the ligand and the metal strengthens. The ligands end up with electrons in their π^* molecular orbital, so the corresponding π bond within the ligand weakens.

The other form of coordination π bonding is ligand-to-metal bonding. This situation arises when the π-symmetry p or π orbitals on the ligands are filled. They combine with the d_{xy}, d_{xz} and d_{yz} orbitals on the metal and donate electrons to the resulting π-symmetry bonding orbital between them and the metal. The metal-ligand bond is somewhat strengthened by this interaction, but the complementary anti-bonding molecular orbital from ligand-to-metal bonding is not higher in energy than the anti-bonding molecular orbital from the σ bonding. It is filled with electrons from the metal d-orbitals, however, becoming the HOMO (highest occupied molecular orbital) of the complex. For that reason, Δ_O decreases when ligand-to-metal bonding occurs.

The greater stabilization that results from metal-to-ligand bonding is caused by the donation of negative charge away from the metal ion, towards the ligands. This allows the metal to accept the σ bonds more easily. The combination of ligand-to-metal σ-bonding and metal-to-ligand π-bonding is a synergic effect, as each enhances the other.

As each of the six ligands has two orbitals of π-symmetry, there are twelve in total. The symmetry adapted linear combinations of these fall into four triply degenerate irreducible representations, one of which is of t_{2g} symmetry. The d_{xy}, d_{xz} and d_{yz} orbitals on the metal also have this symmetry, and so the π-bonds formed between a central metal and six ligands also have it (as these π-bonds are just formed by the overlap of two sets of orbitals with t_{2g} symmetry.)

Role of Metal p-orbitals

Current computational findings suggest valence p orbitals on the metal participate in metal-ligand bonding, albeit weakly. Some new theoretical treatments do not count the metal p-orbitals in metal-ligand bonding, although these orbitals are still included as polarization functions. This model has yet to be adopted by the general chemistry community.

High and Low Spin and the Spectrochemical Series

The six bonding molecular orbitals that are formed are "filled" with the electrons from the ligands, and electrons from the d-orbitals of the metal ion occupy the non-bonding and, in some cases, anti-bonding MOs. The energy difference between the latter two types of MOs is called Δ_O (O stands for octahedral) and is determined by the nature of the π-interaction between the ligand orbitals with the d-orbitals on the central atom. As described above, π-donor ligands lead to a small Δ_O and are called weak- or low-field ligands, whereas π-acceptor ligands lead to a large value of Δ_O and are called strong- or high-field ligands. Ligands that are neither π-donor nor π-acceptor give a value of Δ_O somewhere in-between.

The size of Δ_O determines the electronic structure of the d^4 - d^7 ions. In complexes of metals with these d-electron configurations, the non-bonding and anti-bonding molecular orbitals can be filled in two ways: one in which as many electrons as possible are put in the non-bonding orbitals before filling the anti-bonding orbitals, and one in which as many unpaired electrons as possible are put in. The former case is called low-spin, while the latter is called high-spin. A small Δ_O can be overcome by the energetic gain from not pairing the electrons, leading to high-spin. When Δ_O is large, however, the spin-pairing energy becomes negligible by comparison and a low-spin state arises.

The spectrochemical series is an empirically-derived list of ligands ordered by the size of the splitting Δ that they produce. It can be seen that the low-field ligands are all π-donors (such as I^-), the high field ligands are π-acceptors (such as CN^- and CO), and ligands such as H_2O and NH_3, which are neither, are in the middle.

$I^- < Br^- < S^{2-} < SCN^- < Cl^- < NO_3^- < N_3^- < F^- < OH^- < C_2O_4^{2-} < H_2O < NCS^- < CH_3CN < py$ (pyridine) $< NH_3 < en$ (ethylenediamine) $< bipy$ (2,2'-bipyridine) $< phen$ (1,10-phenanthroline) $< NO_2^- < PPh_3 < CN^- < CO$

Crystal Field Theory

Crystal Field Theory (CFT) is a model that describes the breaking of degeneracies of electron orbital states, usually d or f orbitals, due to a static electric field produced by a surrounding charge distribution (anion neighbors). This theory has been used to describe various spectroscopies of transition metal coordination complexes, in particular optical spectra (colors). CFT successfully accounts for some magnetic properties, colours, hydration enthalpies, and spinel structures of transition metal complexes, but it does not attempt to describe bonding. CFT was developed by physicists Hans Bethe and John Hasbrouck van Vleck in the 1930s. CFT was subsequently combined with molecular orbital theory to form the more realistic and complex ligand field theory (LFT), which delivers insight into the process of chemical bonding in transition metal complexes.

Overview of Crystal Field Theory Analysis

According to Crystal Field Theory, the interaction between a transition metal and ligands arises from the attraction between the positively charged metal cation and negative charge on the non-bonding electrons of the ligand. The theory is developed by considering energy changes of the five degenerate d-orbitals upon being surrounded by an array of point charges consisting of the ligands. As a ligand approaches the metal ion, the electrons from the ligand will be closer to some of the d-orbitals and farther away from others causing a loss of degeneracy. The electrons in the d-orbitals and those in the ligand repel each other due to repulsion between like charges. Thus the d-electrons closer to the ligands will have a higher energy than those further away which results in the d-orbitals splitting in energy. This splitting is affected by the following factors:

- the nature of the metal ion.

- the metal's oxidation state. A higher oxidation state leads to a larger splitting.

- the arrangement of the ligands around the metal ion.

- the nature of the ligands surrounding the metal ion. The stronger the effect of the ligands then the greater the difference between the high and low energy d groups.

The most common type of complex is octahedral; here six ligands form an octahedron around the metal ion. In octahedral symmetry the d-orbitals split into two sets with an energy difference, Δ_{oct} (the crystal-field splitting parameter) where the d_{xy}, d_{xz} and d_{yz} orbitals will be lower in energy than the d_{z^2} and $d_{x^2-y^2}$, which will have higher energy, because the former group is farther from the ligands than the latter and therefore experience less repulsion. The three lower-energy orbitals are collectively referred to as t_{2g}, and the two higher-energy orbitals as e_g. (These labels are based on the theory of molecular symmetry). Typical orbital energy diagrams are given below in the section High-spin and low-spin.

Tetrahedral complexes are the second most common type; here four ligands form a tetrahedron around the metal ion. In a tetrahedral crystal field splitting the d-orbitals again split into two groups, with an energy difference of Δ_{tet} where the lower energy orbitals will be d_{z^2} and $d_{x^2-y^2}$, and the higher energy orbitals will be d_{xy}, d_{xz} and d_{yz} - opposite to the octahedral case. Furthermore, since the ligand electrons in tetrahedral symmetry are not oriented directly towards the d-orbitals, the energy splitting will be lower than in the octahedral case. Square planar and other complex geometries can also be described by CFT.

The size of the gap Δ between the two or more sets of orbitals depends on several factors, including the ligands and geometry of the complex. Some ligands always produce a small value of Δ, while others always give a large splitting. The reasons behind this can be explained by ligand field theory. The spectrochemical series is an empirically-derived list of ligands ordered by the size of the splitting Δ that they produce (small Δ to large Δ; see also this table):

$I^- < Br^- < S^{2-} < SCN^- < Cl^- < NO_3^- < N_3^- < F^- < OH^- < C_2O_4^{2-} < H_2O < NCS^- < CH_3CN <$ py $< NH_3 <$ en $<$ 2,2'-bipyridine $<$ phen $< NO_2^- < PPh_3 < CN^- < CO$

It is useful to note that the ligands producing the most splitting are those that can engage in metal to ligand back-bonding.

The oxidation state of the metal also contributes to the size of Δ between the high and low energy levels. As the oxidation state increases for a given metal, the magnitude of Δ increases. A V^{3+} complex will have a larger Δ than a V^{2+} complex for a given set of ligands, as the difference in charge density allows the ligands to be closer to a V^{3+} ion than to a V^{2+} ion. The smaller distance between the ligand and the metal ion results in a larger Δ, because the ligand and metal electrons are closer together and therefore repel more.

High-spin and Low-spin

Ligands which cause a large splitting Δ of the d-orbitals are referred to as strong-field ligands, such as CN^- and CO from the spectrochemical series. In complexes with these ligands, it is unfavourable to put electrons into the high energy orbitals. Therefore, the lower energy orbitals are completely filled before population of the upper sets starts according to the Aufbau principle. Complexes such as this are called "low spin". For example, NO_2^- is a strong-field ligand and produces a large Δ. The octahedral ion $[Fe(NO_2)_6]^{3-}$, which has 5 d-electrons, would have the octahedral splitting diagram shown at right with all five electrons in the t_{2g} level. The low spin state therefore does not follow Hund's rule.

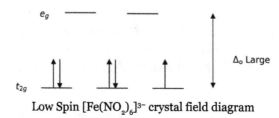

Low Spin $[Fe(NO_2)_6]^{3-}$ crystal field diagram

Conversely, ligands (like I^- and Br^-) which cause a small splitting Δ of the d-orbitals are referred to as weak-field ligands. In this case, it is easier to put electrons into the higher energy set of orbitals than it is to put two into the same low-energy orbital, because two electrons in the same orbital repel each other. So, one electron is put into each of the five d-orbitals before any pairing occurs in accord with Hund's rule and "high spin" complexes are formed. For example, Br^- is a weak-field ligand and produces a small Δ_{oct}. So, the ion $[FeBr_6]^{3-}$, again with five d-electrons, would have an octahedral splitting diagram where all five orbitals are singly occupied.

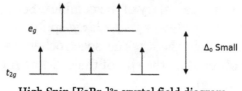

High Spin $[FeBr_6]^{3-}$ crystal field diagram

In order for low spin splitting to occur, the energy cost of placing an electron into an already singly occupied orbital must be less than the cost of placing the additional electron into an e_g orbital at an energy cost of Δ. As noted above, e_g refers to the d_z^2 and $d_{x^2-y^2}$ which are higher in energy than the t_{2g} in octahedral complexes. If the energy required to pair two electrons is greater than the energy cost of placing an electron in an e_g, Δ, high spin splitting occurs.

The crystal field splitting energy for tetrahedral metal complexes (four ligands) is referred to as Δ_{tet}, and is roughly equal to $4/9\Delta_{oct}$ (for the same metal and same ligands). Therefore, the energy required to pair two electrons is typically higher than the energy required for placing electrons in the higher energy orbitals. Thus, tetrahedral complexes are usually high-spin.

The use of these splitting diagrams can aid in the prediction of the magnetic properties of coordination compounds. A compound that has unpaired electrons in its splitting diagram will be paramagnetic and will be attracted by magnetic fields, while a compound that lacks unpaired electrons in its splitting diagram will be diamagnetic and will be weakly repelled by a magnetic field.

Crystal Field Stabilization Energy

The crystal field stabilization energy (CFSE) is the stability that results from placing a transition metal ion in the crystal field generated by a set of ligands. It arises due to the fact that when the d-orbitals are split in a ligand field (as described above), some of them become lower in energy than before with respect to a spherical field known as the barycenter in which all five d-orbitals are degenerate. For example, in an octahedral case, the t_{2g} set becomes lower in energy than the orbitals in the barycenter. As a result of this, if there are any electrons occupying these orbitals, the metal ion is more stable in the ligand field relative to the barycenter by an amount known as the CFSE. Conversely, the e_g orbitals (in the octahedral case) are higher in energy than in the barycenter, so putting electrons in these reduces the amount of CFSE.

If the splitting of the d-orbitals in an octahedral field is Δ_{oct}, the three t_{2g} orbitals are stabilized relative to the barycenter by $2/5\,\Delta_{oct}$, and the e_g orbitals are destabilized by $3/5\,\Delta_{oct}$. As examples, consider the two d^5 configurations shown further up the page. The low-spin (top) example has five electrons in the t_{2g} orbitals, so the total CFSE is $5 \times 2/5\,\Delta_{oct} = 2\Delta_{oct}$. In the high-spin (lower) example, the CFSE is $(3 \times 2/5\,\Delta_{oct}) - (2 \times 3/5\,\Delta_{oct}) = 0$ - in this case, the stabilization generated by the electrons in the lower orbitals is canceled out by the destabilizing effect of the electrons in the upper orbitals.

Crystal Field stabilization is applicable to transition-metal complexes of all geometries. Indeed, the reason that many d^8 complexes are square-planar is the very large amount of crystal field stabilization that this geometry produces with this number of electrons.

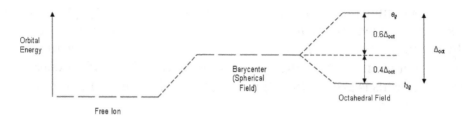

Octahedral crystal field stabilization energy

Explaining the Colors of Transition Metal Complexes

The bright colors exhibited by many coordination compounds can be explained by Crystal Field Theory. If the *d*-orbitals of such a complex have been split into two sets as described above, when the molecule absorbs a photon of visible light one or more electrons may momentarily jump from the lower energy *d*-orbitals to the higher energy ones to transiently create an excited state atom. The difference in energy between the atom in the ground state and in the excited state is equal to the energy of the absorbed photon, and related inversely to the wavelength of the light. Because only certain wavelengths (λ) of light are absorbed - those matching exactly the energy difference - the compounds appears the appropriate complementary color.

As explained above, because different ligands generate crystal fields of different strengths, different colors can be seen. For a given metal ion, weaker field ligands create a complex with a smaller Δ, which will absorb light of longer λ and thus lower frequency v. Conversely, stronger field ligands create a larger Δ, absorb light of shorter λ, and thus higher v. It is, though, rarely the case that the energy of the photon absorbed corresponds exactly to the size of the gap Δ; there are other things (such as electron-electron repulsion and Jahn-Teller effects) that also affect the energy difference between the ground and excited states.

Which Colors are Exhibited?

This color wheel demonstrates which color a compound will appear if it only has one absorption in the visible spectrum. For example, if the compound absorbs red light, it will appear cyan.

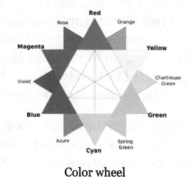

Color wheel

λ absorbed (nm)	Absorbed color	Observed color	Corresponding λ (nm)
400	Violet	Chartreuse yellow	560
450	Blue	Yellow	600
490	Cyan	Red	620
570	Chartreuse green	Violet	410
580	Yellow	Blue	430
600	Orange	Blue	450

Geometries and Crystal Field Splitting Diagrams

Name	Shape	Energy diagram
Octahedral		
Pentagonal bipyramidal		
Square antiprismatic		

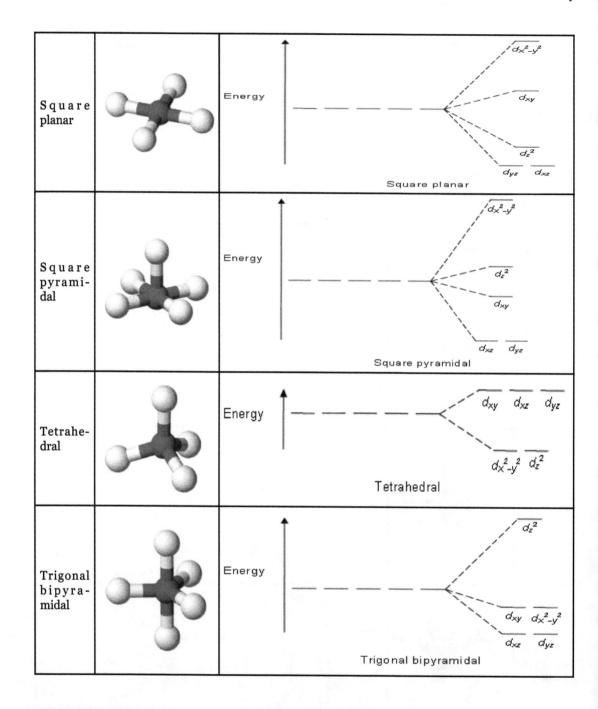

Bohr Model

The Rutherford–Bohr model of the hydrogen atom ($Z = 1$) or a hydrogen-like ion ($Z > 1$), where the negatively charged electron confined to an atomic shell encircles a small, positively charged atomic nucleus and where an electron jump between orbits is accompanied by an emitted or absorbed amount of electromagnetic energy ($h\nu$). The orbits in which the electron may travel are shown as grey circles; their radius increases as n^2, where n is the principal quantum number. The $3 \rightarrow 2$ transition depicted here produces the first line of the Balmer series, and for hydrogen ($Z = 1$) it results in a photon of wavelength 656 nm (red light).

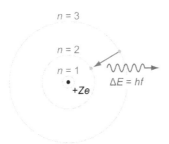

In atomic physics, the Rutherford–Bohr model or Bohr model, introduced by Niels Bohr and Ernest Rutherford in 1913, depicts the atom as a small, positively charged nucleus surrounded by electrons that travel in circular orbits around the nucleus—similar in structure to the Solar System, but with attraction provided by electrostatic forces rather than gravity. After the cubic model (1902), the plum-pudding model (1904), the Saturnian model (1904), and the Rutherford model (1911) came the Rutherford–Bohr model or just *Bohr model* for short (1913). The improvement to the Rutherford model is mostly a quantum physical interpretation of it. The Bohr model has been superseded, but it helped to lead the way to a modern quantum mechanical model of the atom.

The model's key success lay in explaining the Rydberg formula for the spectral emission lines of atomic hydrogen. While the Rydberg formula had been known experimentally, it did not gain a theoretical underpinning until the Bohr model was introduced. Not only did the Bohr model explain the reason for the structure of the Rydberg formula, it also provided a justification for its empirical results in terms of fundamental physical constants.

The Bohr model is a relatively primitive model of the hydrogen atom, compared to the valence shell atom. As a theory, it can be derived as a first-order approximation of the hydrogen atom using the broader and much more accurate quantum mechanics and thus may be considered to be an obsolete scientific theory. However, because of its simplicity, and its correct results for selected systems, the Bohr model is still commonly taught to introduce students to quantum mechanics or energy level diagrams before moving on to the more accurate, but more complex, valence shell atom. A related model was originally proposed by Arthur Erich Haas in 1910, but was rejected. The quantum theory of the period between Planck's discovery of the quantum (1900) and the advent of a full-blown quantum mechanics (1925) is often referred to as the old quantum theory.

Origin

In the early 20th century, experiments by Ernest Rutherford established that atoms consisted of a diffuse cloud of negatively charged electrons surrounding a small, dense, positively charged nucleus. Given this experimental data, Rutherford naturally considered a planetary-model atom, the Rutherford model of 1911 – electrons orbiting a solar nucleus – however, said planetary-model atom has a technical difficulty. The laws of classical mechanics (i.e. the Larmor formula), predict that the electron will release electromagnetic radiation while orbiting a nucleus. Because the electron would lose energy, it would rapidly spiral inwards, collapsing into the nucleus on a timescale of around 16 picoseconds. This atom model is disastrous, because it predicts that all atoms are unstable.

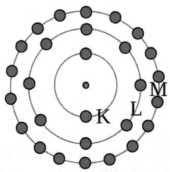

Bohr model showing maximum electrons per shell with shells labeled in X-ray notation

Also, as the electron spirals inward, the emission would rapidly increase in frequency as the orbit got smaller and faster. This would produce a continuous smear, in frequency, of electromagnetic radiation. However, late 19th century experiments with electric discharges have shown that atoms will only emit light (that is, electromagnetic radiation) at certain discrete frequencies.

To overcome this difficulty, Niels Bohr proposed, in 1913, what is now called the *Bohr model of the atom*. He suggested that electrons could only have certain *classical* motions:

1.   Electrons in atoms orbit the nucleus.

2.   The electrons can only orbit stably, without radiating, in certain orbits (called by Bohr the "stationary orbits") at a certain discrete set of distances from the nucleus. These orbits are associated with definite energies and are also called energy shells or energy levels. In these orbits, the electron's acceleration does not result in radiation and energy loss as required by classical electromagnetics. The Bohr model of an atom was based upon Planck's quantum theory of radiation.

3.   Electrons can only gain and lose energy by jumping from one allowed orbit to another, absorbing or emitting electromagnetic radiation with a frequency v determined by the energy difference of the levels according to the Planck relation:

$$\Delta E = E_2 - E_1 = hv \,,$$

where h is Planck's constant. The frequency of the radiation emitted at an orbit of period T is as it would be in classical mechanics; it is the reciprocal of the classical orbit period:

$$v = \frac{1}{T} \; .$$

The significance of the Bohr model is that the laws of classical mechanics apply to the motion of the electron about the nucleus *only when restricted by a quantum rule*. Although Rule 3 is not completely well defined for small orbits, because the emission process involves two orbits with two different periods, Bohr could determine the energy spacing between levels using Rule 3 and come to an exactly correct quantum rule: the angular momentum L is restricted to be an integer multiple of a fixed unit:

$$L = n\frac{h}{2\pi} = n\hbar$$

where $n = 1, 2, 3, \ldots$ is called the principal quantum number, and $\hbar = h/2\pi$. The lowest value of n is 1; this gives a smallest possible orbital radius of 0.0529 nm known as the Bohr radius. Once an electron is in this lowest orbit, it can get no closer to the proton. Starting from the angular momentum quantum rule, Bohr was able to calculate the energies of the allowed orbits of the hydrogen atom and other hydrogen-like atoms and ions.

Other points are:

1. Like Einstein's theory of the Photoelectric effect, Bohr's formula assumes that during a quantum jump a *discrete* amount of energy is radiated. However, unlike Einstein, Bohr stuck to the *classical* Maxwell theory of the electromagnetic field. Quantization of the electromagnetic field was explained by the discreteness of the atomic energy levels; Bohr did not believe in the existence of photons.

2. According to the Maxwell theory the frequency v of classical radiation is equal to the rotation frequency v_{rot} of the electron in its orbit, with harmonics at integer multiples of this frequency. This result is obtained from the Bohr model for jumps between energy levels E_n and E_{n-k} when k is much smaller than n. These jumps reproduce the frequency of the k-th harmonic of orbit n. For sufficiently large values of n (so-called Rydberg states), the two orbits involved in the emission process have nearly the same rotation frequency, so that the classical orbital frequency is not ambiguous. But for small n (or large k), the radiation frequency has no unambiguous classical interpretation. This marks the birth of the correspondence principle, requiring quantum theory to agree with the classical theory only in the limit of large quantum numbers.

3. The Bohr-Kramers-Slater theory (BKS theory) is a failed attempt to extend the Bohr model, which violates the conservation of energy and momentum in quantum jumps, with the conservation laws only holding on average.

Bohr's condition, that the angular momentum is an integer multiple of $\hbar$ was later reinterpreted in 1924 by de Broglie as a standing wave condition: the electron is described by a wave and a whole number of wavelengths must fit along the circumference of the electron's orbit:

$$n\lambda = 2\pi r \ .$$

Substituting de Broglie's wavelength of $\lambda = h/p$ reproduces Bohr's rule. In 1913, however, Bohr justified his rule by appealing to the correspondence principle, without providing any sort of wave interpretation. In 1913, the wave behavior of matter particles such as the electron (i.e., matter waves) was not suspected.

In 1925 a new kind of mechanics was proposed, quantum mechanics, in which Bohr's model of electrons traveling in quantized orbits was extended into a more accurate model of electron motion. The new theory was proposed by Werner Heisenberg. Another form of the same theory, wave mechanics, was discovered by the Austrian physicist Erwin Schrödinger independently, and by different reasoning. Schrödinger employed de Broglie's matter waves, but sought wave solutions of a three-dimensional wave equation describing electrons that were constrained to move about the nucleus of a hydrogen-like atom, by being trapped by the potential of the positive nuclear charge.

Electron Energy Levels

The Bohr model gives almost exact results only for a system where two charged points orbit each other at speeds much less than that of light. This not only includes one-electron systems such as the hydrogen atom, singly ionized helium, doubly ionized lithium, but it includes positronium and Rydberg states of any atom where one electron is far away from everything else. It can be used for K-line X-ray transition calculations if other assumptions are added. In high energy physics, it can be used to calculate the masses of heavy quark mesons.

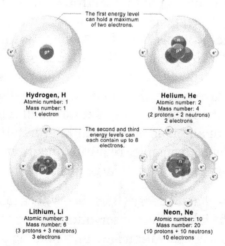

Models depicting electron energy levels in hydrogen, helium, lithium, and neon

Calculation of the orbits requires two assumptions.

- Classical mechanics

    The electron is held in a circular orbit by electrostatic attraction. The centripetal force is equal to the Coulomb force.

    $$\frac{m_e v^2}{r} = \frac{Zk_e e^2}{r^2}$$

    where m_e is the electron's mass, e is the charge of the electron, k_e is Coulomb's constant and Z is the atom's atomic number. It is assumed here that the mass of the nucleus is much larger than the electron mass (which is a good assumption). This equation determines the electron's speed at any radius:

    $$v = \sqrt{\frac{Zk_e e^2}{m_e r}}.$$

    It also determines the electron's total energy at any radius:

    $$E = \frac{1}{2} m_e v^2 - \frac{Zk_e e^2}{r} = -\frac{Zk_e e^2}{2r}.$$

    The total energy is negative and inversely proportional to r. This means that it takes energy to pull the orbiting electron away from the proton. For infinite values of r, the energy is zero, corresponding to a motionless electron infinitely far from the proton. The total energy is half the potential energy, which is also true for noncircular orbits by the virial theorem.

- A quantum rule

    The angular momentum $L = m_e vr$ is an integer multiple of $\hbar$:

    $$m_e vr = n\hbar$$

    Substituting the expression for the velocity gives an equation for r in terms of n:

    $$m_e \sqrt{\frac{k_e Ze^2}{m_e r}}\, r = n\hbar$$

    so that the allowed orbit radius at any n is:

    $$r_n = \frac{n^2 \hbar^2}{Zk_e e^2 m_e}$$

    The smallest possible value of r in the hydrogen atom (Z=1) is called the Bohr radius and is equal to:

$$r_1 = \frac{\hbar^2}{k_e e^2 m_e} \approx 5.29 \times 10^{-11} \, \text{m}$$

The energy of the n-th level for any atom is determined by the radius and quantum number:

$$E = -\frac{Z k_e e^2}{2 r_n} = -\frac{Z^2 (k_e e^2)^2 m_e}{2\hbar^2 n^2} \approx \frac{-13.6 Z^2}{n^2} \, \text{eV}$$

An electron in the lowest energy level of hydrogen ($n = 1$) therefore has about 13.6 eV less energy than a motionless electron infinitely far from the nucleus. The next energy level ($n = 2$) is –3.4 eV. The third ($n = 3$) is –1.51 eV, and so on. For larger values of n, these are also the binding energies of a highly excited atom with one electron in a large circular orbit around the rest of the atom. The hydrogen formula also coincides with the Wallis product.

The combination of natural constants in the energy formula is called the Rydberg energy (R_E):

$$R_E = \frac{(k_e e^2)^2 m_e}{2\hbar^2}$$

This expression is clarified by interpreting it in combinations that form more natural units:

$m_e c^2$ is the rest mass energy of the electron (511 keV)

$\dfrac{k_e e^2}{\hbar c} = \alpha \approx \dfrac{1}{137}$ is the fine structure constant

$R_E = \dfrac{1}{2}(m_e c^2)\alpha^2$

Since this derivation is with the assumption that the nucleus is orbited by one electron, we can generalize this result by letting the nucleus have a charge $q = Z e$ where Z is the atomic number. This will now give us energy levels for hydrogenic atoms, which can serve as a rough order-of-magnitude approximation of the actual energy levels. So for nuclei with Z protons, the energy levels are (to a rough approximation):

$$E_n = -\frac{Z^2 R_E}{n^2}$$

The actual energy levels cannot be solved analytically for more than one electron (see n-body problem) because the electrons are not only affected by the nucleus but also interact with each other via the Coulomb Force.

When $Z = 1/\alpha$ ($Z \approx 137$), the motion becomes highly relativistic, and Z^2 cancels the α^2 in R; the orbit energy begins to be comparable to rest energy. Sufficiently large nuclei, if they were stable, would reduce their charge by creating a bound electron from the vacuum, ejecting the positron to infinity. This is the theoretical phenomenon of electromagnetic charge screening which predicts a maximum nuclear charge. Emission of such positrons has been observed in the collisions of heavy ions to create temporary super-heavy nuclei.

The Bohr formula properly uses the reduced mass of electron and proton in all situations, instead of the mass of the electron,

$$m_{red} = \frac{m_e m_p}{m_e + m_p} = m_e \frac{1}{1 + m_e / m_p} .$$

However, these numbers are very nearly the same, due to the much larger mass of the proton, about 1836.1 times the mass of the electron, so that the reduced mass in the system is the mass of the electron multiplied by the constant 1836.1/(1+1836.1) = 0.99946. This fact was historically important in convincing Rutherford of the importance of Bohr's model, for it explained the fact that the frequencies of lines in the spectra for singly ionized helium do not differ from those of hydrogen by a factor of exactly 4, but rather by 4 times the ratio of the reduced mass for the hydrogen vs. the helium systems, which was much closer to the experimental ratio than exactly 4.

For positronium, the formula uses the reduced mass also, but in this case, it is exactly the electron mass divided by 2. For any value of the radius, the electron and the positron are each moving at half the speed around their common center of mass, and each has only one fourth the kinetic energy. The total kinetic energy is half what it would be for a single electron moving around a heavy nucleus.

$$E_n = \frac{R_E}{2n^2} \quad \text{(positronium)}$$

Rydberg Formula

The Rydberg formula, which was known empirically before Bohr's formula, is seen in Bohr's theory as describing the energies of transitions or quantum jumps between one orbital energy levels. Bohr's formula gives the numerical value of the already-known and measured Rydberg's constant, but in terms of more fundamental constants of nature, including the electron's charge and Planck's constant.

When the electron gets moved from its original energy level to a higher one, it then jumps back each level until it comes to the original position, which results in a photon being emitted. Using the derived formula for the different energy levels of hydrogen one may determine the wavelengths of light that a hydrogen atom can emit.

The energy of a photon emitted by a hydrogen atom is given by the difference of two hydrogen energy levels:

$$E = E_i - E_f = R_E \left(\frac{1}{n_f^2} - \frac{1}{n_i^2} \right)$$

where n_f is the final energy level, and n_i is the initial energy level.

Since the energy of a photon is

$$E = \frac{hc}{\lambda},$$

the wavelength of the photon given off is given by

$$\frac{1}{\lambda} = R \left(\frac{1}{n_f^2} - \frac{1}{n_i^2} \right).$$

This is known as the Rydberg formula, and the Rydberg constant R is R_E/hc, or $R_E/2\pi$ in natural units. This formula was known in the nineteenth century to scientists studying spectroscopy, but there was no theoretical explanation for this form or a theoretical prediction for the value of R, until Bohr. In fact, Bohr's derivation of the Rydberg constant, as well as the concomitant agreement of Bohr's formula with experimentally observed spectral lines of the Lyman (n_f=1), Balmer (n_f=2), and Paschen (n_f=3) series, and successful theoretical prediction of other lines not yet observed, was one reason that his model was immediately accepted.

To apply to atoms with more than one electron, the Rydberg formula can be modified by replacing Z with $Z–b$ or n with $n–b$ where b is constant representing a screening effect due to the inner-shell and other electrons. This was established empirically before Bohr presented his model.

Shell Model of Heavier Atoms

Bohr extended the model of hydrogen to give an approximate model for heavier atoms. This gave a physical picture that reproduced many known atomic properties for the first time.

Heavier atoms have more protons in the nucleus, and more electrons to cancel the charge. Bohr's idea was that each discrete orbit could only hold a certain number of electrons. After that orbit is full, the next level would have to be used. This gives the atom a shell structure, in which each shell corresponds to a Bohr orbit.

This model is even more approximate than the model of hydrogen, because it treats the electrons in each shell as non-interacting. But the repulsions of electrons are taken into account somewhat by the phenomenon of screening. The electrons in outer orbits do not only orbit the nucleus, but they also move around the inner electrons, so the effective charge Z that they feel is reduced by the number of the electrons in the inner orbit.

For example, the lithium atom has two electrons in the lowest 1s orbit, and these orbit at Z=2. Each one sees the nuclear charge of Z=3 minus the screening effect of the other, which crudely reduces the nuclear charge by 1 unit. This means that the innermost electrons orbit at approximately 1/4 the Bohr radius. The outermost electron in lithium orbits at roughly Z=1, since the two inner electrons reduce the nuclear charge by 2. This outer electron should be at nearly one Bohr radius from the nucleus. Because the electrons strongly repel each other, the effective charge description is very approximate; the effective charge Z doesn't usually come out to be an integer. But Moseley's law experimentally probes the innermost pair of electrons, and shows that they do see a nuclear charge of approximately Z–1, while the outermost electron in an atom or ion with only one electron in the outermost shell orbits a core with effective charge Z–k where k is the total number of electrons in the inner shells.

The shell model was able to qualitatively explain many of the mysterious properties of atoms which became codified in the late 19th century in the periodic table of the elements. One property was the size of atoms, which could be determined approximately by measuring the viscosity of gases and density of pure crystalline solids. Atoms tend to get smaller toward the right in the periodic table, and become much larger at the next line of the table. Atoms to the right of the table tend to gain electrons, while atoms to the left tend to lose them. Every element on the last column of the table is chemically inert (noble gas).

In the shell model, this phenomenon is explained by shell-filling. Successive atoms become smaller because they are filling orbits of the same size, until the orbit is full, at which point the next atom in the table has a loosely bound outer electron, causing it to expand. The first Bohr orbit is filled when it has two electrons, which explains why helium is inert. The second orbit allows eight electrons, and when it is full the atom is neon, again inert. The third orbital contains eight again, except that in the more correct Sommerfeld treatment (reproduced in modern quantum mechanics) there are extra "d" electrons. The third orbit may hold an extra 10 d electrons, but these positions are not filled until a few more orbitals from the next level are filled (filling the n=3 d orbitals produces the 10 transition elements). The irregular filling pattern is an effect of interactions between electrons, which are not taken into account in either the Bohr or Sommerfeld models and which are difficult to calculate even in the modern treatment.

Moseley's Law and Calculation of K-alpha X-ray Emission Lines

Niels Bohr said in 1962, "You see actually the Rutherford work [the nuclear atom] was not taken seriously. We cannot understand today, but it was not taken seriously at all. There was no mention of it any place. The great change came from Moseley."

In 1913 Henry Moseley found an empirical relationship between the strongest X-ray line emitted by atoms under electron bombardment (then known as the K-alpha line), and their atomic number Z. Moseley's empiric formula was found to be derivable from Rydberg and Bohr's formula (Moseley actually mentions only Ernest Rutherford and

Antonius Van den Broek in terms of models). The two additional assumptions that this X-ray line came from a transition between energy levels with quantum numbers 1 and 2, and , that the atomic number Z when used in the formula for atoms heavier than hydrogen, should be diminished by 1, to $(Z-1)^2$.

Moseley wrote to Bohr, puzzled about his results, but Bohr was not able to help. At that time, he thought that the postulated innermost "K" shell of electrons should have at least four electrons, not the two which would have neatly explained the result. So Moseley published his results without a theoretical explanation.

Later, people realized that the effect was caused by charge screening, with an inner shell containing only 2 electrons. In the experiment, one of the innermost electrons in the atom is knocked out, leaving a vacancy in the lowest Bohr orbit, which contains a single remaining electron. This vacancy is then filled by an electron from the next orbit, which has n=2. But the n=2 electrons see an effective charge of Z–1, which is the value appropriate for the charge of the nucleus, when a single electron remains in the lowest Bohr orbit to screen the nuclear charge +Z, and lower it by –1 (due to the electron's negative charge screening the nuclear positive charge). The energy gained by an electron dropping from the second shell to the first gives Moseley's law for K-alpha lines,

$$E = hv = E_i - E_f = R_E(Z-1)^2 \left(\frac{1}{1^2} - \frac{1}{2^2} \right)$$

or

$$f = v = R_v \left(\frac{3}{4} \right)(Z-1)^2 = (2.46 \times 10^{15} \text{ Hz})(Z-1)^2.$$

Here, $R_v = R_E/h$ is the Rydberg constant, in terms of frequency equal to 3.28 x 10^{15} Hz. For values of Z between 11 and 31 this latter relationship had been empirically derived by Moseley, in a simple (linear) plot of the square root of X-ray frequency against atomic number (however, for silver, Z = 47, the experimentally obtained screening term should be replaced by 0.4). Notwithstanding its restricted validity, Moseley's law not only established the objective meaning of atomic number but, as Bohr noted, it also did more than the Rydberg derivation to establish the validity of the Rutherford/Van den Broek/Bohr nuclear model of the atom, with atomic number (place on the periodic table) standing for whole units of nuclear charge.

The K-alpha line of Moseley's time is now known to be a pair of close lines, written as ($K\alpha_1$ and $K\alpha_2$) in Siegbahn notation.

Shortcomings

The Bohr model gives an incorrect value $L=\hbar$ for the ground state orbital angular momentum: The angular momentum in the true ground state is known to be zero from experiment. Although mental pictures fail somewhat at these levels of scale, an electron in the lowest modern "orbital" with no orbital momentum, may be thought of as not to rotate "around" the nucleus at all, but merely to go tightly around it in an ellipse with zero area (this may be pictured as "back and forth", without striking or interacting with the nucleus). This is only reproduced in a more sophisticated semiclassical treatment like Sommerfeld's. Still, even the most sophisticated semiclassical model fails to explain the fact that the lowest energy state is spherically symmetric - it doesn't point in any particular direction.

Nevertheless, in the modern *fully quantum treatment in phase space*, the proper deformation (careful full extension) of the semi-classical result adjusts the angular momentum value to the correct effective one. As a consequence, the physical ground state expression is obtained through a shift of the vanishing quantum angular momentum expression, which corresponds to spherical symmetry.

In modern quantum mechanics, the electron in hydrogen is a spherical cloud of probability that grows denser near the nucleus. The rate-constant of probability-decay in hydrogen is equal to the inverse of the Bohr radius, but since Bohr worked with circular orbits, not zero area ellipses, the fact that these two numbers exactly agree is considered a "coincidence". (However, many such coincidental agreements are found between the semiclassical vs. full quantum mechanical treatment of the atom; these include identical energy levels in the hydrogen atom and the derivation of a fine structure constant, which arises from the relativistic Bohr–Sommerfeld model and which happens to be equal to an entirely different concept, in full modern quantum mechanics).

The Bohr model also has difficulty with, or else fails to explain:

- Much of the spectra of larger atoms. At best, it can make predictions about the K-alpha and some L-alpha X-ray emission spectra for larger atoms, if *two* additional ad hoc assumptions are made. Emission spectra for atoms with a single outer-shell electron (atoms in the lithium group) can also be approximately predicted. Also, if the empiric electron–nuclear screening factors for many atoms are known, many other spectral lines can be deduced from the information, in similar atoms of differing elements, via the Ritz–Rydberg combination principles. All these techniques essentially make use of Bohr's Newtonian energy-potential picture of the atom.

- the relative intensities of spectral lines; although in some simple cases, Bohr's formula or modifications of it, was able to provide reasonable estimates (for example, calculations by Kramers for the Stark effect).

- The existence of fine structure and hyperfine structure in spectral lines, which are known to be due to a variety of relativistic and subtle effects, as well as complications from electron spin.

- The Zeeman effect – changes in spectral lines due to external magnetic fields; these are also due to more complicated quantum principles interacting with electron spin and orbital magnetic fields.

- The model also violates the uncertainty principle in that it considers electrons to have known orbits and locations, two things which can not be measured simultaneously.

- Doublets and Triplets: Appear in the spectra of some atoms: Very close pairs of lines. Bohr's model cannot say why some energy levels should be very close together.

- Multi-electron Atoms: don't have energy levels predicted by the model. It doesn't work for (neutral) helium.

- A rotating charge, such as the electron classically orbiting around the nucleus, would constantly lose energy in form of electromagnetic radiation (via various mechanisms: dipole radiation, Bremsstrahlung,...). But such radiation is not observed.

Refinements

Several enhancements to the Bohr model were proposed, most notably the Sommerfeld model or Bohr–Sommerfeld model, which suggested that electrons travel in elliptical orbits around a nucleus instead of the Bohr model's circular orbits. This model supplemented the quantized angular momentum condition of the Bohr model with an additional radial quantization condition, the Sommerfeld–Wilson quantization condition

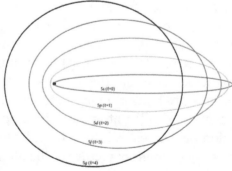

Elliptical orbits with the same energy and quantized angular momentum

$$\int_0^T p_r \, dq_r = nh$$

where p_r is the radial momentum canonically conjugate to the coordinate q which is the radial position and T is one full orbital period. The integral is the action of action-angle coordinates. This condition, suggested by the correspondence principle, is the only one possible, since the quantum numbers are adiabatic invariants.

The Bohr–Sommerfeld model was fundamentally inconsistent and led to many paradoxes. The magnetic quantum number measured the tilt of the orbital plane relative to the xy-plane, and it could only take a few discrete values. This contradicted the obvious fact that an atom could be turned this way and that relative to the coordinates without restriction. The Sommerfeld quantization can be performed in different canonical coordinates and sometimes gives different answers. The incorporation of radiation corrections was difficult, because it required finding action-angle coordinates for a combined radiation/atom system, which is difficult when the radiation is allowed to escape. The whole theory did not extend to non-integrable motions, which meant that many systems could not be treated even in principle. In the end, the model was replaced by the modern quantum mechanical treatment of the hydrogen atom, which was first given by Wolfgang Pauli in 1925, using Heisenberg's matrix mechanics. The current picture of the hydrogen atom is based on the atomic orbitals of wave mechanics which Erwin Schrödinger developed in 1926.

However, this is not to say that the Bohr model was without its successes. Calculations based on the Bohr–Sommerfeld model were able to accurately explain a number of more complex atomic spectral effects. For example, up to first-order perturbations, the Bohr model and quantum mechanics make the same predictions for the spectral line splitting in the Stark effect. At higher-order perturbations, however, the Bohr model and quantum mechanics differ, and measurements of the Stark effect under high field strengths helped confirm the correctness of quantum mechanics over the Bohr model. The prevailing theory behind this difference lies in the shapes of the orbitals of the electrons, which vary according to the energy state of the electron.

The Bohr–Sommerfeld quantization conditions lead to questions in modern mathematics. Consistent semiclassical quantization condition requires a certain type of structure on the phase space, which places topological limitations on the types of symplectic manifolds which can be quantized. In particular, the symplectic form should be the curvature form of a connection of a Hermitian line bundle, which is called a prequantization.

Old Quantum Theory

The old quantum theory is a collection of results from the years 1900–1925 which predate modern quantum mechanics. The theory was never complete or self-consistent, but was a set of heuristic prescriptions which are now understood to be the first quantum corrections to classical mechanics. The Bohr model was the focus of study, and Arnold Sommerfeld made a crucial contribution by quantizing the z-component of the angular momentum, which in the old quantum era was called *space quantization* (Richtungsquantelung). This allowed the orbits of the electron to be ellipses instead of circles, and introduced the concept of quantum degeneracy. The theory would have correctly explained the Zeeman effect, except for the issue of electron spin.

The main tool was Bohr–Sommerfeld quantization, a procedure for selecting out certain discrete set of states of a classical integrable motion as allowed states. These are like the allowed orbits of the Bohr model of the atom; the system can only be in one of these states and not in any states in between.

Basic Principles

The basic idea of the old quantum theory is that the motion in an atomic system is quantized, or discrete. The system obeys classical mechanics except that not every motion is allowed, only those motions which obey the *old quantum condition*:

$$\oint_{H(p,q)=E} p_i \, dq_i = n_i h$$

where the p_i are the momenta of the system and the q_i are the corresponding coordinates. The quantum numbers n_i are *integers* and the integral is taken over one period of the motion at constant energy (as described by the Hamiltonian). The integral is an area in phase space, which is a quantity called the action and is quantized in units of Planck's constant. For this reason, Planck's constant was often called the *quantum of action*.

In order for the old quantum condition to make sense, the classical motion must be separable, meaning that there are separate coordinates q_i in terms of which the motion is periodic. The periods of the different motions do not have to be the same, they can even be incommensurate, but there must be a set of coordinates where the motion decomposes in a multi-periodic way.

The motivation for the old quantum condition was the correspondence principle, complemented by the physical observation that the quantities which are quantized must be adiabatic invariants. Given Planck's quantization rule for the harmonic oscillator, either condition determines the correct classical quantity to quantize in a general system up to an additive constant.

Examples

Thermal Properties of the Harmonic Oscillator

The simplest system in the old quantum theory is the harmonic oscillator, whose Hamiltonian is:

$$H = \frac{p^2}{2m} + \frac{m\omega^2 q^2}{2}.$$

The old quantum theory yields a recipe for the quantization of the energy levels of the harmonic oscillator, which, when combined with the Boltzmann probability distribution of thermodynamics, yields the correct expression for the stored energy and specific heat of a quantum oscillator both at low and at ordinary temperatures. Applied as a model for the specific heat of solids, this resolved a discrepancy in pre-quantum thermodynamics that had troubled 19th-century scientists. Let us now describe this.

The level sets of H are the orbits, and the quantum condition is that the area enclosed by an orbit in phase space is an integer. It follows that the energy is quantized according to the Planck rule:

$$E = n\hbar\omega,$$

a result which was known well before, and used to formulate the old quantum condition. This result differs by $\frac{1}{2}\hbar\omega$ from the results found with the help of quantum mechanics. This constant is neglected in the derivation of the *old quantum theory*, and its value can not be determined using it.

The thermal properties of a quantized oscillator may be found by averaging the energy in each of the discrete states assuming that they are occupied with a Boltzmann weight:

$$U = \frac{\sum_n \hbar\omega n e^{-\beta n\hbar\omega}}{\sum_n e^{-\beta n\hbar\omega}} = \frac{\hbar\omega e^{-\beta\hbar\omega}}{1 - e^{-\beta\hbar\omega}}, \quad \text{where } \beta = \frac{1}{kT},$$

kT is Boltzmann constant times the absolute temperature, which is the temperature as measured in more natural units of energy. The quantity β is more fundamental in thermodynamics than the temperature, because it is the thermodynamic potential associated to the energy.

From this expression, it is easy to see that for large values of β, for very low temperatures, the average energy U in the Harmonic oscillator approaches zero very quickly, exponentially fast. The reason is that kT is the typical energy of random motion at temperature T, and when this is smaller than $\hbar\omega$, there is not enough energy to give the oscillator even one quantum of energy. So the oscillator stays in its ground state, storing next to no energy at all.

This means that at very cold temperatures, the change in energy with respect to beta, or equivalently the change in energy with respect to temperature, is also exponentially small. The change in energy with respect to temperature is the specific heat, so the specific heat is exponentially small at low temperatures, going to zero like

$$\exp(-\hbar\omega / kT)$$

At small values of β, at high temperatures, the average energy U is equal to $1/\beta = kT$. This reproduces the equipartition theorem of classical thermodynamics: every harmonic oscillator at temperature T has energy kT on average. This means that the specific heat of an oscillator is constant in classical mechanics and equal to k. For a collection of atoms connected by springs, a reasonable model of a solid, the total specific heat is equal to the total number of oscillators times k. There are overall three oscillators for each atom, corresponding to the three possible directions of independent oscillations in three dimensions. So the specific heat of a classical solid is always 3k per atom, or in chemistry units, 3R per mole of atoms.

Monatomic solids at room temperatures have approximately the same specific heat of 3k per atom, but at low temperatures they don't. The specific heat is smaller at colder temperatures, and it goes to zero at absolute zero. This is true for all material systems, and this observation is called the third law of thermodynamics. Classical mechanics cannot explain the third law, because in classical mechanics the specific heat is independent of the temperature.

This contradiction between classical mechanics and the specific heat of cold materials was noted by James Clerk Maxwell in the 19th century, and remained a deep puzzle for those who advocated an atomic theory of matter. Einstein resolved this problem in 1906 by proposing that atomic motion is quantized. This was the first application of quantum theory to mechanical systems. A short while later, Peter Debye gave a quantitative theory of solid specific heats in terms of quantized oscillators with various frequencies.

One-dimensional Potential: U=0

One-dimensional problems are easy to solve. At any energy E, the value of the momentum p is found from the conservation equation:

$$\sqrt{2m(E - V(q))} = p$$

which is integrated over all values of q between the classical *turning points*, the places where the momentum vanishes. The integral is easiest for a *particle in a box* of length L, where the quantum condition is:

$$2\int_0^L p\,dq = nh$$

which gives the allowed momenta:

$$p = \frac{nh}{2L}$$

and the energy levels

$$E_n = \frac{p^2}{2m} = \frac{n^2 h^2}{8mL^2}$$

One-dimensional Potential: U=Fx

Another easy case to solve with the old quantum theory is a linear potential on the positive halfline, the constant confining force F binding a particle to an impenetrable wall. This case is much more difficult in the full quantum mechanical treatment, and unlike the other examples, the semiclassical answer here is not exact but approximate, becoming more accurate at large quantum numbers.

$$2\int_0^{\frac{E}{F}} \sqrt{2m(E - Fx)}\, dx = nh$$

so that the quantum condition is

$$\frac{4}{3}\sqrt{2m}\,\frac{E^{3/2}}{F} = nh$$

which determines the energy levels,

$$E_n = \left(\frac{3nhF}{4\sqrt{2m}}\right)^{2/3}$$

In the specific case F=mg, the particle is confined by the gravitational potential of the earth and the "wall" here is the surface of the earth.

One-dimensional Potential: U=(1/2)kx^2

This case is also easy to solve, and the semiclassical answer here agrees with the quantum one to within the ground-state energy. Its quantization-condition integral is

$$2\int_{-\sqrt{\frac{2E}{k}}}^{\sqrt{\frac{2E}{k}}} \sqrt{2m\left(E - \frac{1}{2}kx^2\right)}\, dx = nh$$

with solution

$$E = n\frac{h}{2\pi}\sqrt{\frac{k}{m}} = n\hbar\omega$$

for oscillation angular frequency ω, as before.

Rotator

Another simple system is the rotator. A rotator consists of a mass M at the end of a massless rigid rod of length R and in two dimensions has the Lagrangian:

$$L = \frac{MR^2}{2}\dot{\theta}^2$$

which determines that the angular momentum J conjugate to θ, the polar angle, $J = MR^2\dot{\theta}$. The old quantum condition requires that J multiplied by the period of θ is an integer multiple of Planck's constant:

$$2\pi J = nh$$

the angular momentum to be an integer multiple of $\hbar$. In the Bohr model, this restriction imposed on circular orbits was enough to determine the energy levels.

In three dimensions, a rigid rotator can be described by two angles — θ and ϕ, where θ is the inclination relative to an arbitrarily chosen z-axis while ϕ is the rotator angle in the projection to the x–y plane. The kinetic energy is again the only contribution to the Lagrangian:

$$L = \frac{MR^2}{2}\dot{\theta}^2 + \frac{MR^2}{2}(\sin(\theta)\dot{\phi})^2$$

And the conjugate momenta are $p_\theta = \dot{\theta}$ and $p_\phi = \sin(\theta)^2\dot{\phi}$. The equation of motion for ϕ is trivial: p_ϕ is a constant:

$$p_\phi = l_\phi$$

which is the z-component of the angular momentum. The quantum condition demands that the integral of the constant l_ϕ as ϕ varies from 0 to 2π is an integer multiple of h:

$$l_\phi = m\hbar$$

And m is called the magnetic quantum number, because the z component of the angular momentum is the magnetic moment of the rotator along the z direction in the case where the particle at the end of the rotator is charged.

Since the three-dimensional rotator is rotating about an axis, the total angular momentum should be restricted in the same way as the two-dimensional rotator. The two quantum conditions restrict the total angular momentum and the z-component of the angular momentum to be the integers l,m. This condition is reproduced in modern quantum mechanics, but in the era of the old quantum theory it led to a paradox: how can the orientation of the angular momentum relative to the arbitrarily chosen z-axis be quantized? This seems to pick out a direction in space.

This phenomenon, the quantization of angular momentum about an axis, was given the name *space quantization*, because it seemed incompatible with rotational invariance. In modern quantum mechanics, the angular momentum is quantized the same way, but the discrete states of definite angular momentum in any one orientation are quantum superpositions of the states in other orientations, so that the process of quantization does not pick out a preferred axis. For this reason, the name "space quantization" fell out of favor, and the same phenomenon is now called the quantization of angular momentum.

Hydrogen Atom

The angular part of the hydrogen atom is just the rotator, and gives the quantum numbers l and m. The only remaining variable is the radial coordinate, which executes a periodic one-dimensional potential motion, which can be solved.

For a fixed value of the total angular momentum L, the Hamiltonian for a classical Kepler problem is (the unit of mass and unit of energy redefined to absorb two constants):

$$H = \frac{p^2}{2} + \frac{l^2}{2r^2} - \frac{1}{r}.$$

Fixing the energy to be (a negative) constant and solving for the radial momentum p, the quantum condition integral is:

$$2\oint \sqrt{2E - \frac{l^2}{r^2} + \frac{2}{r}}\, dr = kh$$

which can be solved with the method of residues, and gives a new quantum number k which determines the energy in combination with l. The energy is:

$$E = -\frac{1}{2(k+l)^2}$$

and it only depends on the sum of k and l, which is the *principal quantum number n*. Since k is positive, the allowed values of l for any given n are no bigger than n. The energies reproduce those in the Bohr model, except with the correct quantum mechanical multiplicities, with some ambiguity at the extreme values.

The semiclassical hydrogen atom is called the Sommerfeld model, and its orbits are ellipses of various sizes at discrete inclinations. The Sommerfeld model predicted that the magnetic moment of an atom measured along an axis will only take on discrete values, a result which seems to contradict rotational invariance but which was confirmed by the Stern–Gerlach experiment. This Bohr–Sommerfeld theory is a significant step in the development of quantum mechanics. It also describes the possibility of atomic energy levels being split by a magnetic field (called the Zeeman effect).

Relativistic Orbit

Arnold Sommerfeld derived the relativistic solution of atomic energy levels. We will start this derivation with the relativistic equation for energy in the electric potential

$$W = m_0 c^2 \left(\frac{1}{\sqrt{1 - \dfrac{v^2}{c^2}}} - 1 \right) - k \frac{Ze^2}{r}$$

After substitution $u = \dfrac{1}{r}$ we get

$$\frac{1}{\sqrt{1 - \dfrac{v^2}{c^2}}} = 1 + \frac{W}{m_0 c^2} + k \frac{Ze^2}{m_0 c^2} u$$

For momentum $p_r = m\dot{r}$, $p_\varphi = mr^2 \dot{\varphi}$ and their ratio $\dfrac{p_r}{p_\varphi} = -\dfrac{du}{d\varphi}$ the equation of motion is

$$\frac{d^2 u}{d\varphi^2} = -\left(1 - k^2 \frac{Z^2 e^4}{c^2 p_\varphi^2} \right) u + \frac{m_0 k Z e^2}{p_\varphi^2} \left(1 + \frac{W}{m_0 c^2} \right) = -\omega_0^2 u + K$$

with solution

$$u = \frac{1}{r} = K + A \cos \omega_0 \varphi$$

The angular shift of periapsis per revolution is given by

$$\varphi_s = 2\pi \left(\frac{1}{\omega_0} - 1 \right) \approx 4\pi^3 k^2 \frac{Z^2 e^4}{c^2 n_\varphi^2 h^2}$$

With the quantum conditions

$$\oint p_\varphi \, d\varphi = 2\pi p_\varphi = n_\varphi h$$

and

$$\oint p_r \, dr = p_\varphi \oint \left(\frac{1}{r} \frac{dr}{d\varphi} \right)^2 d\varphi = n_r h$$

we will obtain energies

$$\frac{W}{m_0 c^2} = \left(1 + \frac{\alpha^2 Z^2}{(n_r + \sqrt{n_\varphi^2 - \alpha^2 Z^2})^2}\right)^{-1/2} - 1$$

where α is the fine-structure constant. This solution (using substitutions for quantum numbers) is equivalent to the solution of the Dirac equation. Nevertheless, both solutions fail to predict the Lamb shifts.

De Broglie Waves

In 1905, Einstein noted that the entropy of the quantized electromagnetic field oscillators in a box is, for short wavelength, equal to the entropy of a gas of point particles in the same box. The number of point particles is equal to the number of quanta. Einstein concluded that the quanta could be treated as if they were localizable objects, particles of light, and named them photons.

Einstein's theoretical argument was based on thermodynamics, on counting the number of states, and so was not completely convincing. Nevertheless, he concluded that light had attributes of both waves and particles, more precisely that an electromagnetic standing wave with frequency ω with the quantized energy:

$$E = n\hbar\omega$$

should be thought of as consisting of n photons each with an energy $\hbar\omega$. Einstein could not describe how the photons were related to the wave.

The photons have momentum as well as energy, and the momentum had to be $\hbar k$ where k is the wavenumber of the electromagnetic wave. This is required by relativity, because the momentum and energy form a four-vector, as do the frequency and wave-number.

In 1924, as a PhD candidate, Louis de Broglie proposed a new interpretation of the quantum condition. He suggested that all matter, electrons as well as photons, are described by waves obeying the relations.

$$p = \hbar k$$

or, expressed in terms of wavelength λ instead,

$$p = \frac{h}{\lambda}$$

He then noted that the quantum condition:

$$\int p\,dx = \hbar \int k\,dx = 2\pi\hbar n$$

counts the change in phase for the wave as it travels along the classical orbit, and requires that it be an integer multiple of 2π. Expressed in wavelengths, the number of wavelengths along a classical orbit must be an integer. This is the condition for con-structive interference, and it explained the reason for quantized orbits—the matter waves make standing waves only at discrete frequencies, at discrete energies.

For example, for a particle confined in a box, a standing wave must fit an integer number of wavelengths between twice the distance between the walls. The condition becomes:

$$n\lambda = 2L$$

so that the quantized momenta are:

$$p = \frac{nh}{2L}$$

reproducing the old quantum energy levels.

This development was given a more mathematical form by Einstein, who noted that the phase function for the waves: $\theta(J,x)$ in a mechanical system should be identified with the solution to the Hamilton–Jacobi equation, an equation which even Hamilton considered to be the short-wavelength limit of wave mechanics.

These ideas led to the development of the Schrödinger equation.

Kramers Transition Matrix

The old quantum theory was formulated only for special mechanical systems which could be separated into action angle variables which were periodic. It did not deal with the emission and absorption of radiation. Nevertheless, Hendrik Kramers was able to find heuristics for describing how emission and absorption should be calculated.

Kramers suggested that the orbits of a quantum system should be Fourier analyzed, decomposed into harmonics at multiples of the orbit frequency:

$$X_n(t) = \sum_{k=-\infty}^{\infty} e^{ik\omega t} X_{n;k}$$

The index n describes the quantum numbers of the orbit, it would be $n–l–m$ in the Sommerfeld model. The frequency ω is the angular frequency of the orbit $2\pi / T_n$ while k is an index for the Fourier mode. Bohr had suggested that the k-th harmonic of the classical motion correspond to the transition from level n to level $n–k$.

Kramers proposed that the transition between states were analogous to classical emis-sion of radiation, which happens at frequencies at multiples of the orbit frequencies. The rate of emission of radiation is proportional to $|X_k|^2$, as it would be in classical mechanics. The description was approximate, since the Fourier components did not have frequencies that exactly match the energy spacings between levels.

This idea led to the development of matrix mechanics.

Limitations of The Old Quantum Theory

The old quantum theory had some limitations:

- The old quantum theory provides no means to calculate the intensities of the spectral lines.

- It fails to explain the anomalous Zeeman effect (that is, where the spin of the electron cannot be neglected).

- It cannot quantize "chaotic" systems, i.e. dynamical systems in which trajectories are neither closed nor periodic and whose analytical form does not exist. This presents a problem for systems as simple as a 2-electron atom which is classically chaotic analogously to the famous gravitational three-body problem.

However it can be used to describe atoms with more than one electron (e.g. Helium) and the Zeeman effect. It was later proposed that the old quantum theory is in fact the semi-classical approximation to the canonical quantum mechanics but its limitations are still under investigation.

History

The old quantum theory was sparked by the 1900 work of Max Planck on the emission and absorption of light, and began in earnest after the work of Albert Einstein on the specific heats of solids. Einstein, followed by Debye, applied quantum principles to the motion of atoms, explaining the specific heat anomaly.

In 1913, Niels Bohr identified the correspondence principle and used it to formulate a model of the hydrogen atom which explained the line spectrum. In the next few years Arnold Sommerfeld extended the quantum rule to arbitrary integrable systems making use of the principle of adiabatic invariance of the quantum numbers introduced by Lorentz and Einstein. Sommerfeld's model was much closer to the modern quantum mechanical picture than Bohr's.

Throughout the 1910s and well into the 1920s, many problems were attacked using the old quantum theory with mixed results. Molecular rotation and vibration spectra were understood and the electron's spin was discovered, leading to the confusion of half-integer quantum numbers. Max Planck introduced the zero point energy and Arnold Sommerfeld semiclassically quantized the relativistic hydrogen atom. Hendrik Kramers explained the Stark effect. Bose and Einstein gave the correct quantum statistics for photons.

Kramers gave a prescription for calculating transition probabilities between quantum states in terms of Fourier components of the motion, ideas which were extended in collaboration with Werner Heisenberg to a semiclassical matrix-like description of atomic transition probabilities. Heisenberg went on to reformulate all of quantum theory in terms of a version of these transition matrices, creating matrix mechanics.

In 1924, Louis de Broglie introduced the wave theory of matter, which was extended to a semiclassical equation for matter waves by Albert Einstein a short time later. In 1926 Erwin Schrödinger found a completely quantum mechanical wave-equation, which reproduced all the successes of the old quantum theory without ambiguities and inconsistencies. Schrödinger's wave mechanics developed separately from matrix mechanics until Schrödinger and others proved that the two methods predicted the same experimental consequences. Paul Dirac later proved in 1926 that both methods can be obtained from a more general method called transformation theory.

References

- Petrucci, R. H.; W. S., Harwood; F. G., Herring (2002). General Chemistry: Principles and Modern Applications (8th ed.). Prentice-Hall. p. 413–414 (Table 11.1). ISBN 0-13-014329-4.

- Miessler, G. L.; Tarr, D. A. (1999). Inorganic Chemistry (2nd ed.). Prentice-Hall. p. 54–62. ISBN 0-13-841891-8.

- Hanson, Robert M. (1995). Molecular origami: precision scale models from paper. University Science Books. ISBN 0-935702-30-X.

- Greenwood, Norman N.; Earnshaw, Alan (1997). Chemistry of the Elements (2nd ed.). Butterworth-Heinemann. ISBN 0-08-037941-9.

- Murrel, J.N.; Kettle, S.F.A.; Tedder, J.M. (1985). The Chemical Bond (2nd ed.). John Wiley & Sons. ISBN 0-471-90759-6.

- Shaik, Sason S.; Phillipe C. Hiberty (2008). A Chemist's Guide to Valence Bond Theory. New Jersey: Wiley-Interscience. ISBN 978-0-470-03735-5.

- Miller, Bernard (2004). Advanced Organic Chemistry: Reactions and Mechanisms. Upper Saddle River, NJ: Pearsons. pp. 53–54. ISBN 0-13-065588-0.

- Frenking, Gernot; Shaik, Sason, eds. (May 2014). "Chapter 7: Chemical bonding in Transition Metal Compounds". The Chemical Bond: Chemical Bonding Across the Periodic Table. Wiley-VCH. ISBN 978-3-527-33315-8.

- L.D. Landau, E.M. Lifshitz (1977). Quantum Mechanics: Non-Relativistic Theory. Vol. 3 (3rd ed.). Pergamon Press. ISBN 978-0-08-020940-1.

Orbitals and Hydrogen Atoms

Atomic orbitals are the areas of the atom where the probability of finding electrons is high and these orbitals are designated by three exclusive quantum numbers n, l and m. Molecular orbital on the other hand describes the chemical or physical properties of an electron in a molecule. In the chapter the reader is introduced to molecular and atomic orbitals and the molecular orbital theory. There is a section on the hydrogen atom and systems like muonium, positronium etc. that can be categorized as hydrogen-like atoms.

Atomic Orbital

In quantum mechanics, an atomic orbital is a mathematical function that describes the wave-like behavior of either one electron or a pair of electrons in an atom. This function can be used to calculate the probability of finding any electron of an atom in any specific region around the atom's nucleus. The term, atomic orbital, may also refer to the physical region or space where the electron can be calculated to be present, as defined by the particular mathematical form of the orbital.

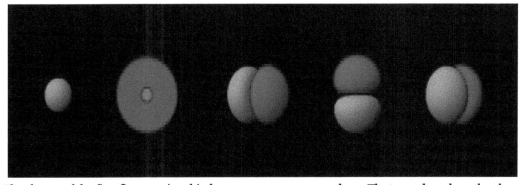

The shapes of the first five atomic orbitals are: 1s, 2s, $2p_x$, $2p_y$, and $2p_z$. The two colors show the phase or sign of the wave function in each region. These are graphs of $\psi(x, y, z)$ functions which depend on the coordinates of one electron.

Each orbital in an atom is characterized by a unique set of values of the three quantum numbers n, ℓ, and m, which respectively correspond to the electron's energy, angular momentum, and an angular momentum vector component (the magnetic quantum number). Each such orbital can be occupied by a maximum of two electrons, each with its own spin quantum number s. The simple names s orbital, p orbital, d orbital and f orbital refer to orbitals with angular momentum quantum number ℓ = 0, 1, 2 and 3 respectively. These names, together with the value of n, are used to describe the electron configurations of atoms. They are derived from the description by early spectroscopists of certain series of alkali metal spectroscopic lines as sharp, principal, diffuse, and fundamental. Orbitals for $\ell > 3$ continue alphabetically, omitting j (g, h, i, k, ...) because some languages do not distinguish between the letters "i" and "j":

Atomic orbitals are the basic building blocks of the atomic orbital model (alternatively known as the electron cloud or wave mechanics model), a modern framework for visualizing the submicroscopic behavior of electrons in matter. In this model the electron cloud of a multi-electron atom may be seen as being built up (in approximation) in an electron configuration that is a product of simpler hydrogen-like atomic orbitals. The repeating *periodicity* of the blocks of 2, 6, 10, and 14 elements within sections of the periodic table arises naturally from the total number of electrons that occupy a complete set of s, p, d and f atomic orbitals, respectively, although for higher values of the quantum number n, particularly when the atom in question bears a positive charge, the energies of certain sub-shells become very similar and so the order in which they are said to be populated by electrons (e.g. $Cr = [Ar]4s^1 3d^5$ and $Cr^{2+} = [Ar]3d^4$) can only be rationalized somewhat arbitrarily.

Electron Properties

With the development of quantum mechanics and experimental findings (such as the two slits diffraction of electrons), it was found that the orbiting electrons around a nucleus could not be fully described as particles, but needed to be explained by the wave-particle duality. In this sense, the electrons have the following properties:

Wave-like properties:

1. The electrons do not orbit the nucleus in the manner of a planet orbiting the sun, but instead exist as standing waves. The lowest possible energy an electron can take is therefore analogous to the fundamental frequency of a wave on a string. Higher energy states are then similar to harmonics of the fundamental frequency.

2. The electrons are never in a single point location, although the probability of interacting with the electron at a single point can be found from the wave function of the electron.

Particle-like properties:

1. There is always an integer number of electrons orbiting the nucleus.

2. Electrons jump between orbitals in a particle-like fashion. For example, if a single photon strikes the electrons, only a single electron changes states in response to the photon.

3. The electrons retain particle-like properties such as: each wave state has the same electrical charge as the electron particle. Each wave state has a single discrete spin (spin up or spin down). This can depend upon its superposition.

Thus, despite the popular analogy to planets revolving around the Sun, electrons cannot be described simply as solid particles. In addition, atomic orbitals do not closely resemble a planet's elliptical path in ordinary atoms. A more accurate analogy might be that of a large and often oddly shaped "atmosphere" (the electron), distributed around a relatively tiny planet (the atomic nucleus). Atomic orbitals exactly describe the shape of this "atmosphere" only when a single electron is present in an atom. When more electrons are added to a single atom, the additional electrons tend to more evenly fill in a volume of space around the nucleus so that the resulting collection (sometimes termed the atom's "electron cloud") tends toward a generally spherical zone of probability describing where the atom's electrons will be found. The is due to the uncertainty principle.

Formal Quantum Mechanical Definition

Atomic orbitals may be defined more precisely in formal quantum mechanical language. Specifically, in quantum mechanics, the state of an atom, i.e., an eigenstate of the atomic Hamiltonian, is approximated by an expansion into linear combinations of anti-symmetrized products (Slater determinants) of one-electron functions. The spatial components of these one-electron functions are called atomic orbitals. (When one considers also their spin component, one speaks of atomic spin orbitals.) A state is actually a function of the coordinates of all the electrons, so that their motion is correlated, but this is often approximated by this independent-particle model of products of single electron wave functions. (The London dispersion force, for example, depends on the correlations of the motion of the electrons.)

In atomic physics, the atomic spectral lines correspond to transitions (quantum leaps) between quantum states of an atom. These states are labeled by a set of quantum numbers summarized in the term symbol and usually associated with particular electron configurations, i.e., by occupation schemes of atomic orbitals (for example, $1s^2\ 2s^2\ 2p^6$ for the ground state of neon—term symbol: $^1S_0$).

This notation means that the corresponding Slater determinants have a clear higher weight in the configuration interaction expansion. The atomic orbital concept is therefore a key concept for visualizing the excitation process associated with a given transition. For example, one can say for a given transition that it corresponds to the excitation of an electron from an occupied orbital to a given unoccupied orbital. Nevertheless, one has to keep in mind that electrons are fermions ruled by the Pauli exclusion principle and cannot be distinguished from the other electrons in the atom. Moreover, it sometimes happens that the configuration interaction expansion converges very slowly and that one cannot speak about simple one-determinant wave function at all. This is the case when electron correlation is large.

Fundamentally, an atomic orbital is a one-electron wave function, even though most electrons do not exist in one-electron atoms, and so the one-electron view is an approximation. When thinking about orbitals, we are often given an orbital vision which (even if it is not spelled out) is heavily influenced by this Hartree–Fock approximation, which is one way to reduce the complexities of molecular orbital theory.

Types of Orbitals

Atomic orbitals can be the hydrogen-like "orbitals" which are exact solutions to the Schrödinger equation for a hydrogen-like "atom" (i.e., an atom with one electron). Alternatively, atomic orbitals refer to functions that depend on the coordinates of one electron (i.e., orbitals) but are used as starting points for approximating wave functions that depend on the simultaneous coordinates of all the electrons in an atom or molecule. The coordinate systems chosen for atomic orbitals are usually spherical coordinates (r, θ, φ) in atoms and cartesians (x, y, z) in polyatomic molecules. The advantage of spherical coordinates (for atoms) is that an orbital wave function is a product of three factors each dependent on a single coordinate: $\psi(r, \theta, \varphi) = R(r)\,\Theta(\theta)\,\Phi(\varphi)$. The angular factors of atomic orbitals $\Theta(\theta)\,\Phi(\varphi)$ generate s, p, d, etc. functions as real combinations of spherical harmonics $Y_{\ell m}(\theta, \varphi)$ (where ℓ and m are quantum numbers). There are typically three mathematical forms for the radial functions $R(r)$ which can be chosen as a starting point for the calculation of the properties of atoms and molecules with many electrons:

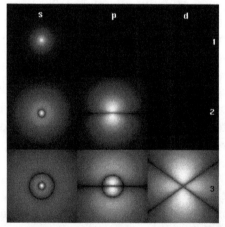

False-color density images of some hydrogen-like atomic orbitals (f orbitals and higher are not shown)

1. The *hydrogen-like atomic orbitals* are derived from the exact solution of the Schrödinger Equation for one electron and a nucleus, for a hydrogen-like atom. The part of the function that depends on the distance from the nucleus has nodes (radial nodes) and decays as e.

2. The Slater-type orbital (STO) is a form without radial nodes but decays from the nucleus as does the hydrogen-like orbital.

3. The form of the Gaussian type orbital (Gaussians) has no radial nodes and decays as e.

Although hydrogen-like orbitals are still used as pedagogical tools, the advent of computers has made STOs preferable for atoms and diatomic molecules since combinations of STOs can replace the nodes in hydrogen-like atomic orbital. Gaussians are typically used in molecules with three or more atoms. Although not as accurate by themselves as STOs, combinations of many Gaussians can attain the accuracy of hydrogen-like orbitals.

History

The term "orbital" was coined by Robert Mulliken in 1932 as an abbreviation for *one-electron orbital wave function*. However, the idea that electrons might revolve around a compact nucleus with definite angular momentum was convincingly argued at least 19 years earlier by Niels Bohr, and the Japanese physicist Hantaro Nagaoka published an orbit-based hypothesis for electronic behavior as early as 1904. Explaining the behavior of these electron "orbits" was one of the driving forces behind the development of quantum mechanics.

Early Models

With J.J. Thomson's discovery of the electron in 1897, it became clear that atoms were not the smallest building blocks of nature, but were rather composite particles. The newly discovered structure within atoms tempted many to imagine how the atom's constituent parts might interact with each other. Thomson theorized that multiple electrons revolved in orbit-like rings within a positively charged jelly-like substance, and between the electron's discovery and 1909, this "plum pudding model" was the most widely accepted explanation of atomic structure.

Shortly after Thomson's discovery, Hantaro Nagaoka predicted a different model for electronic structure. Unlike the plum pudding model, the positive charge in Nagaoka's "Saturnian Model" was concentrated into a central core, pulling the electrons into circular orbits reminiscent of Saturn's rings. Few people took notice of Nagaoka's work at the time, and Nagaoka himself recognized a fundamental defect in the theory even at its conception, namely that a classical charged object cannot sustain orbital motion because it is accelerating and therefore loses energy due to electromagnetic radiation. Nevertheless, the Saturnian model turned out to have more in common with modern theory than any of its contemporaries.

Bohr Atom

In 1909, Ernest Rutherford discovered that the bulk of the atomic mass was tightly condensed into a nucleus, which was also found to be positively charged. It became clear from his analysis in 1911 that the plum pudding model could not explain atomic structure. In 1913 as Rutherford's post-doctoral student, Niels Bohr proposed a new model of the atom, wherein electrons orbited the nucleus with classical periods, but were only permitted to have discrete values of angular momentum, quantized in units $h/2\pi$. This constraint automatically permitted only certain values of electron energies. The Bohr model of the atom fixed the problem of energy loss from radiation from a ground state (by declaring that there was no state below this), and more importantly explained the origin of spectral lines.

After Bohr's use of Einstein's explanation of the photoelectric effect to relate energy levels in atoms with the wavelength of emitted light, the connection between the structure of electrons in atoms and the emission and absorption spectra of atoms became an increasingly useful tool in the understanding of electrons in atoms. The most prominent feature of emission and absorption spectra (known experimentally since the middle of the 19th century), was that these atomic spectra contained discrete lines. The significance of the Bohr model was that it related the lines in emission and absorption spectra to the energy differences between the orbits that electrons could take around an atom. This was, however, *not* achieved by Bohr through giving the electrons some kind of wave-like properties, since the idea that electrons could behave as matter waves was not suggested until eleven years later. Still, the Bohr model's use of quantized angular momenta and therefore quantized energy levels was a significant step towards the understanding of electrons in atoms, and also a significant step towards the development of quantum mechanics in suggesting that quantized restraints must account for all discontinuous energy levels and spectra in atoms.

With de Broglie's suggestion of the existence of electron matter waves in 1924, and for a short time before the full 1926 Schrödinger equation treatment of hydrogen-like atom, a Bohr electron "wavelength" could be seen to be a function of its momentum, and thus a Bohr orbiting electron was seen to orbit in a circle at a multiple of its half-wavelength (this physically incorrect Bohr model is still often taught to beginning students). The Bohr model for a short time could be seen as a classical model with an additional constraint provided by the 'wavelength' argument. However, this period was immediately superseded by the full three-dimensional wave mechanics of 1926. In our current understanding of physics, the Bohr model is called a semi-classical model because of its quantization of angular momentum, not primarily because of its relationship with electron wavelength, which appeared in hindsight a dozen years after the Bohr model was proposed.

The Bohr model was able to explain the emission and absorption spectra of hydrogen. The energies of electrons in the $n = 1, 2, 3$, etc. states in the Bohr model match those of current physics. However, this did not explain similarities between different atoms, as expressed by the periodic table, such as the fact that helium (two electrons), neon (10 electrons), and argon (18 electrons) exhibit similar chemical inertness. Modern quantum mechanics explains this in terms of electron shells and subshells which can each hold a number of electrons determined by the Pauli exclusion principle. Thus the $n = 1$ state can hold one or two electrons, while the $n = 2$ state can hold up to eight electrons in 2s and 2p subshells. In helium, all $n = 1$ states are fully occupied; the same for $n = 1$ and $n = 2$ in neon. In argon the 3s and 3p subshells are similarly fully occupied by eight electrons; quantum mechanics also allows a 3d subshell but this is at higher energy than the 3s and 3p in argon (contrary to the situation in the hydrogen atom) and remains empty.

Modern Conceptions and Connections to the Heisenberg Uncertainty Principle

Immediately after Heisenberg discovered his uncertainty principle, it was noted by Bohr that the existence of any sort of wave packet implies uncertainty in the wave frequency and wavelength, since a spread of frequencies is needed to create the packet itself. In quantum mechanics, where all particle momenta are associated with waves, it is the formation of such a wave packet which localizes the wave, and thus the particle, in space. In states where a quantum mechanical particle is bound, it must be localized as a wave packet, and the existence of the packet and its minimum size implies a spread and minimal value in particle wavelength, and thus also momentum and energy. In quantum mechanics, as a particle is localized to a smaller region in space, the associated compressed wave packet requires a larger and larger range of momenta, and thus larger kinetic energy. Thus, the binding energy to contain or trap a particle in a smaller region of space, increases without bound, as the region of space grows smaller. Particles cannot be restricted to a geometric point in space, since this would require an infinite particle momentum.

In chemistry, Schrödinger, Pauling, Mulliken and others noted that the consequence of Heisenberg's relation was that the electron, as a wave packet, could not be considered to have an exact location in its orbital. Max Born suggested that the electron's position needed to be described by a probability distribution which was connected with finding the electron at some point in the wave-function which described its associated wave packet. The new quantum mechanics did not give exact results, but only the probabilities for the occurrence of a variety of possible such results. Heisenberg held that the path of a moving particle has no meaning if we cannot observe it, as we cannot with electrons in an atom.

In the quantum picture of Heisenberg, Schrödinger and others, the Bohr atom number n for each orbital became known as an *n-sphere* in a three dimensional atom and was pictured as the mean energy of the probability cloud of the electron's wave packet which surrounded the atom.

Orbital Names

Orbitals are given names in the form:

$$X \text{ type}^y$$

where X is the energy level corresponding to the principal quantum number n, type is a lower-case letter denoting the shape or subshell of the orbital and it corresponds to the angular quantum number ℓ, and y is the number of electrons in that orbital.

For example, the orbital $1s^2$ (pronounced "one ess two") has two electrons and is the lowest energy level ($n = 1$) and has an angular quantum number of $\ell = 0$. In X-ray notation, the principal quantum number is given a letter associated with it. For $n = 1, 2, 3, 4, 5, ...$, the letters associated with those numbers are K, L, M, N, O, ... respectively.

Hydrogen-like Orbitals

The simplest atomic orbitals are those that are calculated for systems with a single electron, such as the hydrogen atom. An atom of any other element ionized down to a single electron is very similar to hydrogen, and the orbitals take the same form. In the Schrödinger equation for this system of one negative and one positive particle, the atomic orbitals are the eigenstates of the Hamiltonian operator for the energy. They can be obtained analytically, meaning that the resulting orbitals are products of a poly-nomial series, and exponential and trigonometric functions.

For atoms with two or more electrons, the governing equations can only be solved with the use of methods of iterative approximation. Orbitals of multi-electron atoms are *qualitatively* similar to those of hydrogen, and in the simplest models, they are taken to have the same form. For more rigorous and precise analysis, the numerical approximations must be used.

A given (hydrogen-like) atomic orbital is identified by unique values of three quantum numbers: n, ℓ, and m_ℓ. The rules restricting the values of the quantum numbers, and their energies, explain the electron configuration of the atoms and the periodic table.

The stationary states (quantum states) of the hydrogen-like atoms are its atomic orbitals. However, in general, an electron's behavior is not fully described by a single orbital. Electron states are best represented by time-depending "mixtures" (linear combinations) of multiple orbitals.

The quantum number n first appeared in the Bohr model where it determines the radius of each circular electron orbit. In modern quantum mechanics however, n determines the mean distance of the electron from the nucleus; all electrons with the same value of n lie at the same average distance. For this reason, orbitals with the same value of n are said to comprise a "shell". Orbitals with the same value of n and also the same value of ℓ are even more closely related, and are said to comprise a "subshell".

Quantum Numbers

Because of the quantum mechanical nature of the electrons around a nucleus, atomic orbitals can be uniquely defined by a set of integers known as quantum numbers. These quantum numbers only occur in certain combinations of values, and their physical interpretation changes depending on whether real or complex versions of the atomic orbitals are employed.

Complex Orbitals

In physics, the most common orbital descriptions are based on the solutions to the hydrogen atom, where orbitals are given by the product between a radial function and a pure spherical harmonic. The quantum numbers, together with the rules governing their possible values, are as follows:

The principal quantum number n describes the energy of the electron and is always a positive integer. In fact, it can be any positive integer, but for reasons discussed below, large numbers are seldom encountered. Each atom has, in general, many orbitals associated with each value of n; these orbitals together are sometimes called *electron shells*.

The azimuthal quantum number ℓ describes the orbital angular momentum of each electron and is a non-negative integer. Within a shell where n is some integer n_0, ℓ ranges across all (integer) values satisfying the relation $0 \leq \ell \leq n_0 - 1$. For instance, the $n = 1$ shell has only orbitals with $\ell = 0$, and the $n = 2$ shell has only orbitals with $\ell = 0$, and $\ell = 1$. The set of orbitals associated with a particular value of ℓ are sometimes collectively called a *subshell*.

The magnetic quantum number, m_ℓ, describes the magnetic moment of an electron in an arbitrary direction, and is also always an integer. Within a subshell where ℓ is some integer ℓ_0, m_ℓ ranges thus: $-\ell_0 \leq m_\ell \leq \ell_0$.

The above results may be summarized in the following table. Each cell represents a subshell, and lists the values of m_ℓ available in that subshell. Empty cells represent subshells that do not exist.

	$\ell = 0$	$\ell = 1$	$\ell = 2$	$\ell = 3$	$\ell = 4$	...
$n = 1$	$m\ell = 0$					
$n = 2$	0	−1, 0, 1				
$n = 3$	0	−1, 0, 1	−2, −1, 0, 1, 2			
$n = 4$	0	−1, 0, 1	−2, −1, 0, 1, 2	−3, −2, −1, 0, 1, 2, 3		
$n = 5$	0	−1, 0, 1	−2, −1, 0, 1, 2	−3, −2, −1, 0, 1, 2, 3	−4, −3, −2, −1, 0, 1, 2, 3, 4	
...	...	...	...	...	...	...

Subshells are usually identified by their n- and ℓ-values. n is represented by its numerical value, but ℓ is represented by a letter as follows: 0 is represented by 's', 1 by 'p', 2 by 'd', 3 by 'f', and 4 by 'g'. For instance, one may speak of the subshell with $n = 2$ and $\ell = 0$ as a '2s subshell'.

Each electron also has a spin quantum number, **s**, which describes the spin of each electron (spin up or spin down). The number **s** can be +1/2 or −1/2.

The Pauli exclusion principle states that no two electrons can occupy the same quantum state: every electron in an atom must have a unique combination of quantum numbers.

The above conventions imply a preferred axis (for example, the z direction in Cartesian coordinates), and they also imply a preferred direction along this preferred axis. Otherwise there would be no sense in distinguishing $m = +1$ from $m = -1$. As such, the model is most useful when applied to physical systems that share these symmetries. The Stern–Gerlach experiment — where an atom is exposed to a magnetic field — provides one such example.

Real Orbitals

An atom that is embedded in a crystalline solid feels multiple preferred axes, but no preferred direction. Instead of building atomic orbitals out of the product of radial functions and a single spherical harmonic, linear combinations of spherical harmonics are typically used, designed so that the imaginary part of the spherical harmonics cancel out. These real orbitals are the building blocks most commonly shown in orbital visualizations.

In the real hydrogen-like orbitals, for example, n and ℓ have the same interpretation and significance as their complex counterparts, but m is no longer a good quantum number (though its absolute value is). The orbitals are given new names based on their shape with respect to a standardized Cartesian basis. The real hydrogen-like p orbitals are given by the following

$$p_z = p_0$$
$$p_x = \frac{1}{\sqrt{2}}\left(p_1 + p_{-1}\right)$$

$$p_y = \frac{1}{i\sqrt{2}}\left(p_1 - p_{-1}\right)$$

where $p_0 = R_{n1}Y_{10}$, $p_1 = R_{n1}Y_{11}$, and $p_{-1} = R_{n1}Y_{1-1}$, are the complex orbitals corresponding to $\ell = 1$.

The equations for the p_x and p_y orbitals depend on the phase convention used for the spherical harmonics. The above equations suppose that the spherical harmonics are defined by $Y_\ell^m(\theta, \varphi) = Ne^{im\varphi}P_\ell^m(\cos\theta)$. However some quantum physicists include a phase factor $(-1)^m$ in these definitions, which has the effect of relating the p_x orbital to a *difference* of spherical harmonics and the p_y orbital to the corresponding *sum*. (For more detail, see Spherical harmonics#Conventions).

Shapes of Orbitals

Simple pictures showing orbital shapes are intended to describe the angular forms of regions in space where the electrons occupying the orbital are likely to be found. The diagrams cannot, however, show the entire region where an electron can be found, since according to quantum mechanics there is a non-zero probability of finding the electron (almost) anywhere in space. Instead the diagrams are approximate representations of boundary or contour surfaces where the probability density $|\psi(r, \theta, \varphi)|^2$ has a constant value, chosen so that there is a certain probability (for example 90%) of finding the electron within the contour. Although $|\psi|^2$ as the square of an absolute value is everywhere non-negative, the sign of the wave function $\psi(r, \theta, \varphi)$ is often indicated in each subregion of the orbital picture.

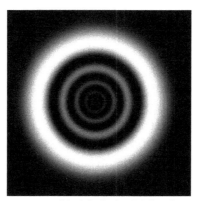

Cross-section of computed hydrogen atom orbital ($\psi(r, \theta, \varphi)^2$) for the 6s ($n = 6, \ell = 0, m = 0$) orbital. Note that s orbitals, though spherically symmetrical, have radially placed wave-nodes for $n > 1$. However, only s orbitals invariably have a center anti-node; the other types never do.

Sometimes the ψ function will be graphed to show its phases, rather than the $|\psi(r, \theta, \varphi)|^2$ which shows probability density but has no phases (which have been lost in the process of taking the absolute value, since $\psi(r, \theta, \varphi)$ is a complex number). $|\psi(r, \theta, \varphi)|^2$ orbital graphs tend to have less spherical, thinner lobes than $\psi(r, \theta, \varphi)$ graphs, but have the same number of lobes in the same places, and otherwise are recognizable. This article, in order to show wave function phases, shows mostly $\psi(r, \theta, \varphi)$ graphs.

The lobes can be viewed as standing wave interference patterns between the two counter rotating, ring resonant travelling wave "m" and "$-m$" modes, with the projection of the orbital onto the xy plane having a resonant "m" wavelengths around the circumference. Though rarely depicted the travelling wave solutions can be viewed as rotating banded tori, with the bands representing phase information. For each m there are two standing wave solutions $\langle m \rangle + \langle -m \rangle$ and $\langle m \rangle - \langle -m \rangle$. For the case where $m = 0$ the orbital is vertical, counter rotating information is unknown, and the orbital is z-axis symmetric. For the case where $\ell = 0$ there are no counter rotating modes. There are only radial modes and the shape is spherically symmetric. For any given n, the smaller ℓ is, the more radial nodes there are. Loosely speaking n is energy, ℓ is analogous to eccentricity, and m is orientation. For the record, in the classical case, a ring resonant travelling wave, for example in a circular transmission line, unless actively forced, will spontaneously decay into a ring resonant standing wave because reflections will build up over time at even the smallest imperfection or discontinuity.

Generally speaking, the number n determines the size and energy of the orbital for a given nucleus: as n increases, the size of the orbital increases. However, in comparing different elements, the higher nuclear charge Z of heavier elements causes their orbitals to contract by comparison to lighter ones, so that the overall size of the whole atom remains very roughly constant, even as the number of electrons in heavier elements (higher Z) increases.

Also in general terms, ℓ determines an orbital's shape, and m_ℓ its orientation. However, since some orbitals are described by equations in complex numbers, the shape sometimes depends on m_ℓ also. Together, the whole set of orbitals for a given ℓ and n fill space as symmetrically as possible, though with increasingly complex sets of lobes and nodes.

The single s-orbitals ($\ell = 0$) are shaped like spheres. For $n = 1$ it is roughly a solid ball (it is most dense at the center and fades exponentially outwardly), but for $n = 2$ or more, each single s-orbital is composed of spherically symmetric surfaces which are nested shells (i.e., the "wave-structure" is radial, following a sinusoidal radial component as well). See illustration of a cross-section of these nested shells, at right. The s-orbitals for all n numbers are the only orbitals with an anti-node (a region of high wave func-tion density) at the center of the nucleus. All other orbitals (p, d, f, etc.) have angular momentum, and thus avoid the nucleus (having a wave node *at* the nucleus). Recently, there has been an effort to experimentally image the 1s and 2p orbitals in a $SrTiO_3$ crystal using scanning transmission electron microscopy with energy dispersive x-ray spectroscopy. Because the imaging was conducted using an electron beam, Coulombic beam-orbital interaction that is often termed as the impact parameter effect is included in the final outcome.

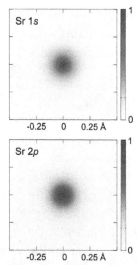

Experimentally imaged 1s and 2p core-electron orbitals of Sr, including the effects of atomic thermal vibrations and excitation broadening, retrieved from energy dispersive x-ray spectroscopy (EDX) in scanning transmission electron microscopy (STEM).

The shapes of p, d and f-orbitals are described verbally here and shown graphically in the *Orbitals table* below. The three p-orbitals for $n = 2$ have the form of two ellipsoids with a point of tangency at the nucleus (the two-lobed shape is sometimes referred to as a "dumbbell"—there are two lobes pointing in opposite directions from each other). The three p-orbitals in each shell are oriented at right angles to each other, as determined by their respective linear combination of values of m_t. The overall result is a lobe pointing along each direction of the primary axes.

Four of the five d-orbitals for $n = 3$ look similar, each with four pear-shaped lobes, each lobe tangent at right angles to two others, and the centers of all four lying in one plane. Three of these planes are the xy-, xz-, and yz-planes—the lobes are between the pairs of primary axes—and the fourth has the centres along the x and y axes themselves. The fifth and final d-orbital consists of three regions of high probability density: a torus with two pear-shaped regions placed symmetrically on its z axis. The overall total of 18 directional lobes point in every primary axis direction and between every pair.

There are seven f-orbitals, each with shapes more complex than those of the d-orbitals.

Additionally, as is the case with the s orbitals, individual p, d, f and g orbitals with n values higher than the lowest possible value, exhibit an additional radial node structure which is reminiscent of harmonic waves of the same type, as compared with the lowest (or fundamental) mode of the wave. As with s orbitals, this phenomenon provides p, d, f, and g orbitals at the next higher possible value of n (for example, 3p orbitals vs. the fundamental 2p), an additional node in each lobe. Still higher values of n further increase the number of radial nodes, for each type of orbital.

The shapes of atomic orbitals in one-electron atom are related to 3-dimensional spherical harmonics. These shapes are not unique, and any linear combination is valid, like a transformation to cubic harmonics, in fact it is possible to generate sets where all the d's are the same shape, just like the p_x, p_y, and p_z are the same shape.

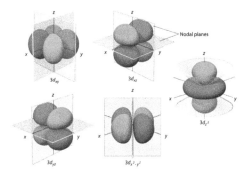

Qualitative Understanding of Shapes

The shapes of atomic orbitals can be understood qualitatively by considering the analogous case of standing waves on a circular drum. To see the analogy, the mean vibrational displacement of each bit of drum membrane from the equilibrium point over many cycles (a measure of average drum membrane velocity and momentum at that point) must be considered relative to that point's distance from the center of the drum head. If this displacement is taken as being analogous to the probability of finding an electron at a given distance from the nucleus, then it will be seen that the many modes of the vibrating disk form patterns that trace the various shapes of atomic orbitals. The basic reason for this correspondence lies in the fact that the distribution of kinetic energy and momentum in a matter-wave is predictive of where the particle associated with the wave will be. That is, the probability of finding an electron at a given place is also a function of the electron's average momentum at that point, since high electron momentum at a given position tends to "localize" the electron in that position, via the properties of electron wave-packets.

This relationship means that certain key features can be observed in both drum membrane modes and atomic orbitals. For example, in all of the modes analogous to **s** orbitals (the top row in the animated illustration below), it can be seen that the very center of the drum membrane vibrates most strongly, corresponding to the antinode in all **s** orbitals in an atom. This antinode means the electron is most likely to be at the physical position of the nucleus (which it passes straight through without scattering or striking it), since it is moving (on average) most rapidly at that point, giving it maximal momentum.

A mental "planetary orbit" picture closest to the behavior of electrons in **s** orbitals, all of which have no angular momentum, might perhaps be that of a Keplerian orbit with the orbital eccentricity of 1 but a finite major axis, not physically possible (because particles were to collide), but can be imagined as a limit of orbits with equal major axes but increasing eccentricity.

Below, a number of drum membrane vibration modes and the respective wave functions of the hydrogen atom are shown. A correspondence can be considered where the wave functions of a vibrating drum head are for a two-coordinate system $\psi(r, \theta)$ and the wave functions for a vibrating sphere are three-coordinate $\psi(r, \theta, \varphi)$.

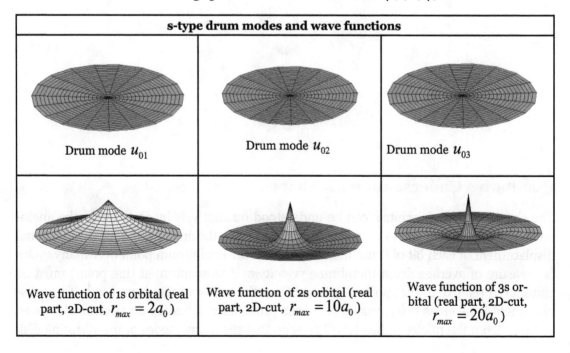

s-type drum modes and wave functions		
Drum mode u_{01}	Drum mode u_{02}	Drum mode u_{03}
Wave function of 1s orbital (real part, 2D-cut, $r_{max} = 2a_0$)	Wave function of 2s orbital (real part, 2D-cut, $r_{max} = 10a_0$)	Wave function of 3s orbital (real part, 2D-cut, $r_{max} = 20a_0$)

None of the other sets of modes in a drum membrane have a central antinode, and in all of them the center of the drum does not move. These correspond to a node at the nucleus for all non-**s** orbitals in an atom. These orbitals all have some angular momentum, and in the planetary model, they correspond to particles in orbit with eccentricity less than 1.0, so that they do not pass straight through the center of the primary body, but keep somewhat away from it.

In addition, the drum modes analogous to p and d modes in an atom show spatial irregularity along the different radial directions from the center of the drum, whereas all of the modes analogous to **s** modes are perfectly symmetrical in radial direction. The non radial-symmetry properties of non-**s** orbitals are necessary to localize a particle with angular momentum and a wave nature in an orbital where it must tend to stay away from the central attraction force, since any particle localized at the point of central attraction could have no angular momentum. For these modes, waves in the drum head tend to avoid the central point. Such features again emphasize that the shapes of atomic orbitals are a direct consequence of the wave nature of electrons.

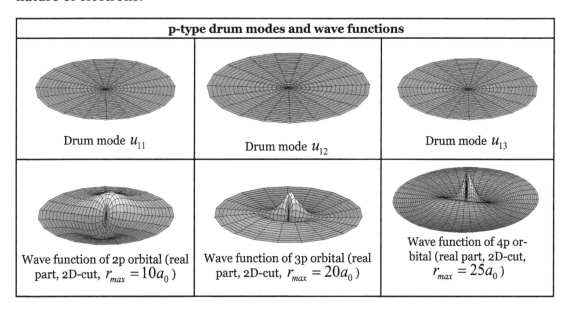

p-type drum modes and wave functions		
Drum mode u_{11}	Drum mode u_{12}	Drum mode u_{13}
Wave function of 2p orbital (real part, 2D-cut, $r_{max} = 10a_0$)	Wave function of 3p orbital (real part, 2D-cut, $r_{max} = 20a_0$)	Wave function of 4p orbital (real part, 2D-cut, $r_{max} = 25a_0$)

d-type Drum Modes

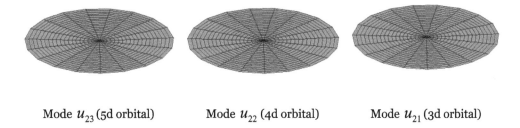

Mode u_{23} (5d orbital)   Mode u_{22} (4d orbital)   Mode u_{21} (3d orbital)

Orbital Energy

In atoms with a single electron (hydrogen-like atoms), the energy of an orbital (and, consequently, of any electrons in the orbital) is determined exclusively by n. The $n = 1$ orbital has the lowest possible energy in the atom. Each successively higher value of n has a higher level of energy, but the difference decreases as n increases. For high n, the level of energy becomes so high that the electron can easily escape from the atom. In single electron atoms, all levels with different ℓ within a given n are (to a good approximation) degenerate, and have the same energy. This approximation is broken to a slight extent by the effect of the magnetic field of the nucleus, and by quantum electrodynamics effects. The latter induce tiny binding energy differences especially for **s** electrons that go nearer the nucleus, since these feel a very slightly different nuclear charge, even in one-electron atoms.

In atoms with multiple electrons, the energy of an electron depends not only on the intrinsic properties of its orbital, but also on its interactions with the other electrons. These interactions depend on the detail of its spatial probability distribution, and so the energy levels of orbitals depend not only on n but also on ℓ. Higher values of ℓ are associated with higher values of energy; for instance, the 2p state is higher than the 2s state. When $\ell = 2$, the increase in energy of the orbital becomes so large as to push the energy of orbital above the energy of the s-orbital in the next higher shell; when $\ell = 3$ the energy is pushed into the shell two steps higher. The filling of the 3d orbitals does not occur until the 4s orbitals have been filled.

The increase in energy for subshells of increasing angular momentum in larger atoms is due to electron–electron interaction effects, and it is specifically related to the ability of low angular momentum electrons to penetrate more effectively toward the nucleus, where they are subject to less screening from the charge of intervening electrons. Thus, in atoms of higher atomic number, the ℓ of electrons becomes more and more of a determining factor in their energy, and the principal quantum numbers n of electrons becomes less and less important in their energy placement.

The energy sequence of the first 35 subshells (e.g., 1s, 2p, 3d, etc.) is given in the following table. Each cell represents a subshell with n and ℓ given by its row and column indices, respectively. The number in the cell is the subshell's position in the sequence. For a linear listing of the subshells in terms of increasing energies in multielectron atoms, see the section below.

	s	p	d	f	g	h
1	1					
2	2	3				
3	4	5	7			
4	6	8	10	13		
5	9	11	14	17	*21*	

6	12	15	18	22	26	31
7	16	19	23	27	32	37
8	20	24	28	33	38	44
9	25	29	34	39	45	51
10	30	35	40	46	52	59

Note: empty cells indicate non-existent sublevels, while numbers in italics indicate sublevels that could (potentially) exist, but which do not hold electrons in any element currently known.

Electron Placement and the Periodic Table

Several rules govern the placement of electrons in orbitals (*electron configuration*). The first dictates that no two electrons in an atom may have the same set of values of quantum numbers (this is the Pauli exclusion principle). These quantum numbers include the three that define orbitals, as well as *s*, or spin quantum number. Thus, two electrons may occupy a single orbital, so long as they have different values of *s*. However-er, *only* two electrons, because of their spin, can be associated with each orbital.

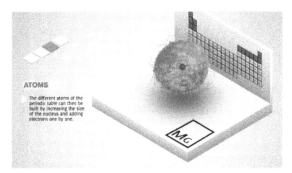

ATOMS

The different atoms of the periodic table can then be built by increasing the size of the nucleus and adding electrons one by one.

Atomic orbitals and periodic table constructionT

Additionally, an electron always tends to fall to the lowest possible energy state. It is possible for it to occupy any orbital so long as it does not violate the Pauli exclusion principle, but if lower-energy orbitals are available, this condition is unstable. The electron will eventually lose energy (by releasing a photon) and drop into the lower orbital. Thus, electrons fill orbitals in the order specified by the energy sequence given above.

This behavior is responsible for the structure of the periodic table. The table may be divided into several rows (called 'periods'), numbered starting with 1 at the top. The presently known elements occupy seven periods. If a certain period has number *i*, it consists of elements whose outermost electrons fall in the *i*th shell. Niels Bohr was the first to propose (1923) that the periodicity in the properties of the elements might be explained by the periodic filling of the electron energy levels, resulting in the electronic structure of the atom.

The periodic table may also be divided into several numbered rectangular 'blocks'. The elements belonging to a given block have this common feature: their highest-energy electrons all belong to the same ℓ-state (but the n associated with that ℓ-state depends upon the period). For instance, the leftmost two columns constitute the 's-block'. The outermost electrons of Li and Be respectively belong to the 2s subshell, and those of Na and Mg to the 3s subshell.

The following is the order for filling the "subshell" orbitals, which also gives the order of the "blocks" in the periodic table:

1s, 2s, 2p, 3s, 3p, 4s, 3d, 4p, 5s, 4d, 5p, 6s, 4f, 5d, 6p, 7s, 5f, 6d, 7p

						1s								
2s												2p	2p	2p
3s												3p	3p	3p
4s							3d	3d	3d	3d	3d	4p	4p	4p
5s							4d	4d	4d	4d	4d	5p	5p	5p
6s	4f	4f	4f	4f	4f	4f	4f	5d	5d	5d	5d	5d	6p	6p	6p
7s	5f	5f	5f	5f	5f	5f	5f	6d	6d	6d	6d	6d	7p	7p	7p

The "periodic" nature of the filling of orbitals, as well as emergence of the **s**, p, d and f "blocks", is more obvious if this order of filling is given in matrix form, with increasing principal quantum numbers starting the new rows ("periods") in the matrix. Then, each subshell (composed of the first two quantum numbers) is repeated as many times as required for each pair of electrons it may contain. The result is a compressed periodic table, with each entry representing two successive elements:

Although this is the general order of orbital filling according to the Madelung rule, there are exceptions, and the actual electronic energies of each element are also depen-dent upon additional details of the atoms.

The number of electrons in an electrically neutral atom increases with the atomic number. The electrons in the outermost shell, or *valence electrons*, tend to be responsible for an element's chemical behavior. Elements that contain the same number of valence electrons can be grouped together and display similar chemical properties.

Relativistic Effects

For elements with high atomic number Z, the effects of relativity become more pronounced, and especially so for s electrons, which move at relativistic velocities as they penetrate the screening electrons near the core of high-Z atoms. This relativistic increase in momentum for high speed electrons causes a corresponding decrease in wavelength and contraction of 6s orbitals relative to 5d orbitals (by comparison to corresponding s and d electrons in lighter elements in the same column of the periodic table); this results in 6s valence electrons becoming lowered in energy.

Examples of significant physical outcomes of this effect include the lowered melting temperature of mercury (which results from 6s electrons not being available for metal bonding) and the golden color of gold and caesium (which results from narrowing of 6s to 5d transition energy to the point that visible light begins to be absorbed).

In the Bohr Model, an $n = 1$ electron has a velocity given by $v = Z\alpha c$, where Z is the atomic number, α is the fine-structure constant, and c is the speed of light. In non-relativistic quantum mechanics, therefore, any atom with an atomic number greater than 137 would require its 1s electrons to be traveling faster than the speed of light. Even in the Dirac equation, which accounts for relativistic effects, the wave function of the electron for atoms with $Z > 137$ is oscillatory and unbounded. The significance of element 137, also known as untriseptium, was first pointed out by the physicist Richard Feynman. Element 137 is sometimes informally called feynmanium (symbol Fy). However, Feynman's approximation fails to predict the exact critical value of Z due to the non-point-charge nature of the nucleus and very small orbital radius of inner electrons, resulting in a potential seen by inner electrons which is effectively less than Z. The critical Z value which makes the atom unstable with regard to high-field breakdown of the vacuum and production of electron-positron pairs, does not occur until Z is about 173. These conditions are not seen except transiently in collisions of very heavy nuclei such as lead or uranium in accelerators, where such electron-positron production from these effects has been claimed to be observed.

There are no nodes in relativistic orbital densities, although individual components of the wave function will have nodes.

Transitions Between Orbitals

Bound quantum states have discrete energy levels. When applied to atomic orbitals, this means that the energy differences between states are also discrete. A transition between these states (i.e., an electron absorbing or emitting a photon) can thus only happen if the photon has an energy corresponding with the exact energy difference between said states.

Consider two states of the hydrogen atom:

State 1) $n = 1$, $\ell = 0$, $m_\ell = 0$ and $s = +1/2$

State 2) $n = 2$, $\ell = 0$, $m_\ell = 0$ and $s = +1/2$

By quantum theory, state 1 has a fixed energy of E_1, and state 2 has a fixed energy of

E_2. Now, what would happen if an electron in state 1 were to move to state 2? For this to happen, the electron would need to gain an energy of exactly $E_2 - E_1$. If the electron receives energy that is less than or greater than this value, it cannot jump from state 1 to state 2. Now, suppose we irradiate the atom with a broad-spectrum of light. Photons that reach the atom that have an energy of exactly $E_2 - E_1$ will be absorbed by the electron in state 1, and that electron will jump to state 2. However, photons that are greater or lower in energy cannot be absorbed by the electron, because the electron can only jump to one of the orbitals, it cannot jump to a state between orbitals. The result is that only photons of a specific frequency will be absorbed by the atom. This creates a line in the spectrum, known as an absorption line, which corresponds to the energy difference between states 1 and 2.

The atomic orbital model thus predicts line spectra, which are observed experimentally. This is one of the main validations of the atomic orbital model.

The atomic orbital model is nevertheless an approximation to the full quantum theory, which only recognizes many electron states. The predictions of line spectra are qualitatively useful but are not quantitatively accurate for atoms and ions other than those containing only one electron.

Molecular Orbital

In chemistry, a molecular orbital (MO) is a mathematical function describing the wave-like behavior of an electron in a molecule. This function can be used to calculate chemical and physical properties such as the probability of finding an electron in any specific region. The term *orbital* was introduced by Robert S. Mulliken in 1932 as an abbreviation for *one-electron orbital wave function*. At an elementary level, it is used to describe the *region* of space in which the function has a significant amplitude. Molecular orbitals are usually constructed by combining atomic orbitals or hybrid orbitals from each atom of the molecule, or other molecular orbitals from groups of atoms. They can be quantitatively calculated using the Hartree–Fock or self-consistent field (SCF) methods.

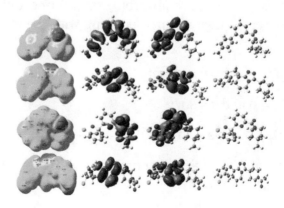

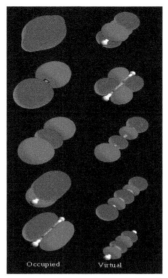

Complete acetylene (H–C≡C–H) molecular orbital set. The left column shows MO's which are occupied in the ground state, with the lowest-energy orbital at the top. The white and grey line visible in some MO's is the molecular axis passing through the nuclei. The orbital wave functions are positive in the red regions and negative in the blue. The right column shows virtual MO's which are empty in the ground state, but may be occupied in excited states.

Overview

A molecular orbital (MO) can be used to represent the regions in a molecule where an electron occupying that orbital is likely to be found. Molecular orbitals are obtained from the combination of atomic orbitals, which predict the location of an electron in an atom. A molecular orbital can specify the electron configuration of a molecule: the spatial distribution and energy of one (or one pair of) electron(s). Most commonly a MO is represented as a linear combination of atomic orbitals (the LCAO-MO method), especially in qualitative or very approximate usage. They are invaluable in providing a simple model of bonding in molecules, understood through molecular orbital theory. Most present-day methods in computational chemistry begin by calculating the MOs of the system. A molecular orbital describes the behavior of one electron in the electric field generated by the nuclei and some average distribution of the other electrons. In the case of two electrons occupying the same orbital, the Pauli principle demands that they have opposite spin. Necessarily this is an approximation, and highly accurate de-scriptions of the molecular electronic wave function do not have orbitals.

Formation of Molecular Orbitals

Molecular orbitals arise from allowed interactions between atomic orbitals, which are allowed if the symmetries (determined from group theory) of the atomic or-bitals are compatible with each other. Efficiency of atomic orbital interactions is determined from the overlap (a measure of how well two orbitals constructively interact with one another) between two atomic orbitals, which is significant if the atomic orbitals are close in energy. Finally, the number of molecular orbitals that form must equal the number of atomic orbitals in the atoms being combined to form the molecule.

Qualitative Discussion

For an imprecise, but qualitatively useful, discussion of the molecular structure, the molecular orbitals can be obtained from the "Linear combination of atomic orbitals molecular orbital method" ansatz. Here, the molecular orbitals are expressed as linear combinations of atomic orbitals.

Linear Combinations of Atomic Orbitals (LCAO)

Molecular orbitals were first introduced by Friedrich Hund and Robert S. Mulliken in 1927 and 1928. The linear combination of atomic orbitals or "LCAO" approximation for molecular orbitals was introduced in 1929 by Sir John Lennard-Jones. His ground-breaking paper showed how to derive the electronic structure of the fluorine and oxygen molecules from quantum principles. This qualitative approach to molecular orbital theory is part of the start of modern quantum chemistry. Linear combinations of atomic orbitals (LCAO) can be used to estimate the molecular orbitals that are formed upon bonding between the molecule's constituent atoms. Similar to an atomic orbital, a Schrödinger equation, which describes the behavior of an electron, can be constructed for a molecular orbital as well. Linear combinations of atomic orbitals, or the sums and differences of the atomic wavefunctions, provide approximate solutions to the Hartree–Fock equations which correspond to the independent-particle approximation of the molecular Schrödinger equation. For simple diatomic molecules, the wavefunctions obtained are represented mathematically by the equations

$$\Psi = c_a \psi_a + c_b \psi_b$$

$$\Psi^* = c_a \psi_a - c_b \psi_b$$

where Ψ and Ψ^* are the molecular wavefunctions for the bonding and antibonding molecular orbitals, respectively, ψ_a and ψ_b are the atomic wavefunctions from atoms a and b, respectively, and c_a and c_b are adjustable coefficients. These coefficients can be positive or negative, depending on the energies and symmetries of the individual atomic orbitals. As the two atoms become closer together, their atomic orbitals overlap to produce areas of high electron density, and, as a consequence, molecular orbitals are formed between the two atoms. The atoms are held together by the electrostatic attraction between the positively charged nuclei and the negatively charged electrons occupying bonding molecular orbitals.

Bonding, Antibonding, and Nonbonding MOs

When atomic orbitals interact, the resulting molecular orbital can be of three types: bonding, antibonding, or nonbonding.

Bonding MOs:

- Bonding interactions between atomic orbitals are constructive (in-phase) interactions.

- Bonding MOs are lower in energy than the atomic orbitals that combine to produce them.

Antibonding MOs:

- Antibonding interactions between atomic orbitals are destructive (out-of-phase) interactions, with a nodal plane where the wavefunction of the antibonding orbital is zero between the two interacting atoms

- Antibonding MOs are higher in energy than the atomic orbitals that combine to produce them.

Nonbonding MOs:

- Nonbonding MOs are the result of no interaction between atomic orbitals because of lack of compatible symmetries.

- Nonbonding MOs will have the same energy as the atomic orbitals of one of the atoms in the molecule.

Sigma and Pi Labels for MOs

The type of interaction between atomic orbitals can be further categorized by the molecular-orbital symmetry labels σ (sigma), π (pi), δ (delta), φ (phi), γ (gamma) etc. paralleling the symmetry of the atomic orbitals s, p, d, f and g. The number of nodal planes containing the internuclear axis between the atoms concerned is zero for σ MOs, one for π, two for δ, etc.

σ Symmetry

A MO with σ symmetry results from the interaction of either two atomic s-orbitals or two atomic p_z-orbitals. An MO will have σ-symmetry if the orbital is symmetric with respect to the axis joining the two nuclear centers, the internuclear axis. This means that rotation of the MO about the internuclear axis does not result in a phase change. A σ^* orbital, sigma antibonding orbital, also maintains the same phase when rotated about the internuclear axis. The σ^* orbital has a nodal plane that is between the nuclei and perpendicular to the internuclear axis.

π Symmetry

A MO with π symmetry results from the interaction of either two atomic p_x orbitals or p_y orbitals. An MO will have π symmetry if the orbital is asymmetric with respect to rotation about the internuclear axis. This means that rotation of the MO about the internuclear axis will result in a phase change. There is one nodal plane containing the internuclear axis, if real orbitals are considered.

A π* orbital, pi antibonding orbital, will also produce a phase change when rotated about the internuclear axis. The π* orbital also has a second nodal plane between the nuclei.

δ Symmetry

A MO with δ symmetry results from the interaction of two atomic d_{xy} or $d_{x^2-y^2}$ orbitals. Because these molecular orbitals involve low-energy d atomic orbitals, they are seen in transition-metal complexes. A δ bonding orbital has two nodal planes containing the internuclear axis, and a δ* antibonding orbital also has a third nodal plane between the nuclei.

φ Symmetry

Theoretical chemists have conjectured that higher-order bonds, such as phi bonds corresponding to overlap of f atomic orbitals, are possible. There is as of 2005 only one known example of a molecule purported to contain a phi bond (a U–U bond, in the molecule U_2).

Suitably aligned f atomic orbitals overlap to form phi molecular orbital (a phi bond)

Gerade and Ungerade Symmetry

For molecules that possess a center of inversion (centrosymmetric molecules) there are additional labels of symmetry that can be applied to molecular orbitals. Centrosymmetric molecules include:

- Homonuclear diatomics, X_2

- Octahedral, EX_6

- Square planar, EX_4.

Non-centrosymmetric molecules include:

- Heteronuclear diatomics, XY

- Tetrahedral, EX_4.

If inversion through the center of symmetry in a molecule results in the same phases for the molecular orbital, then the MO is said to have gerade (g) symmetry, from the German word for even. If inversion through the center of symmetry in a molecule results in a phase change for the molecular orbital, then the MO is said to have ungerade (u) symmetry, from the German word for odd. For a bonding MO with σ-symmetry, the orbital is σ_g (s' + s" is symmetric), while an antibonding MO with σ-symmetry the orbital is σ_u, because inversion of s' – s" is antisymmetric. For a bonding MO with π-symmetry the orbital is π_u because inversion through the center of symmetry for would produce a sign change (the two p atomic orbitals are in phase with each other but the two lobes have opposite signs), while an antibonding MO with π-symmetry is π_g because inversion through the center of symmetry for would not produce a sign change (the two p orbitals are antisymmetric by phase).

MO Diagrams

The qualitative approach of MO analysis uses a molecular orbital diagram to visualize bonding interactions in a molecule. In this type of diagram, the molecular orbitals are represented by horizontal lines; the higher a line the higher the energy of the orbital, and degenerate orbitals are placed on the same level with a space between them. Then, the electrons to be placed in the molecular orbitals are slotted in one by one, keeping in mind the Pauli exclusion principle and Hund's rule of maximum multiplicity (only 2 electrons, having opposite spins, per orbital; place as many unpaired electrons on one energy level as possible before starting to pair them). For more complicated molecules, the wave mechanics approach loses utility in a qualitative understanding of bonding (although is still necessary for a quantitative approach). Some properties:

- A basis set of orbitals includes those atomic orbitals that are available for molecular orbital interactions, which may be bonding or antibonding

- The number of molecular orbitals is equal to the number of atomic orbitals included in the linear expansion or the basis set

- If the molecule has some symmetry, the degenerate atomic orbitals (with the same atomic energy) are grouped in linear combinations (called **symmetry-adapted atomic orbitals (SO)**), which belong to the representation of the symmetry group, so the wave functions that describe the group are known as symmetry-adapted linear combinations (SALC).

- The number of molecular orbitals belonging to one group representation is equal to the number of symmetry-adapted atomic orbitals belonging to this representation

- Within a particular representation, the symmetry-adapted atomic orbitals mix more if their atomic energy levels are closer.

The general procedure for constructing a molecular orbital diagram for a reasonably simple molecule can be summarized as follows:

1.   Assign a point group to the molecule.

2.   Look up the shapes of the SALCs.

3.   Arrange the SALCs of each molecular fragment in increasing order of energy, first noting whether they stem from s, p, or d orbitals (and put them in the order $s < p < d$), and then their number of internuclear nodes.

4.   Combine SALCs of the same symmetry type from the two fragments, and from N SALCs form N molecular orbitals.

5.   Estimate the relative energies of the molecular orbitals from considerations of overlap and relative energies of the parent orbitals, and draw the levels on a molecular orbital energy level diagram (showing the origin of the orbitals).

6.   Confirm, correct, and revise this qualitative order by carrying out a molecular orbital calculation by using commercial software.

Bonding in Molecular Orbitals

Orbital Degeneracy

Molecular orbitals are said to be degenerate if they have the same energy. For example, in the homonuclear diatomic molecules of the first ten elements, the molecular orbitals derived from the p_x and the p_y atomic orbitals result in two degenerate bonding orbitals (of low energy) and two degenerate antibonding orbitals (of high energy).

Ionic Bonds

When the energy difference between the atomic orbitals of two atoms is quite large, one atom's orbitals contribute almost entirely to the bonding orbitals, and the others atom's orbitals contribute almost entirely to the antibonding orbitals. Thus, the situation is effectively that some electrons have been transferred from one atom to the other. This is called an (mostly) ionic bond.

Bond Order

The bond order, or number of bonds, of a molecule can be determined by combining the number of electrons in bonding and antibonding molecular orbitals. A pair of electrons in a bonding orbital creates a bond, whereas a pair of electrons in an antibonding orbital negates a bond. For example, N_2, with eight electrons in bonding orbitals and two electrons in antibonding orbitals, has a bond order of three, which constitutes a triple bond.

Bond strength is proportional to bond order—a greater amount of bonding produces a more stable bond—and bond length is inversely proportional to it—a stronger bond is shorter.

There are rare exceptions to the requirement of molecule having a positive bond order. Although Be_2 has a bond order of 0 according to MO analysis, there is experimental evidence of a highly unstable Be_2 molecule having a bond length of 245 pm and bond energy of 10 kJ/mol.

HOMO and LUMO

The highest occupied molecular orbital and lowest unoccupied molecular orbital are often referred to as the HOMO and LUMO, respectively. The difference of the energies of the HOMO and LUMO, termed the band gap, can sometimes serve as a measure of the excitability of the molecule: The smaller the energy the more easily it will be excited.

Molecular Orbital Examples

Homonuclear Diatomics

Homonuclear diatomic MOs contain equal contributions from each atomic orbital in the basis set. This is shown in the homonuclear diatomic MO diagrams for H_2, He_2, and Li_2, all of which containing symmetric orbitals.

H_2

As a simple MO example consider the hydrogen molecule, H_2 (see molecular orbital diagram), with the two atoms labelled H' and H". The lowest-energy atomic orbitals, 1s' and 1s", do not transform according to the symmetries of the molecule.

The symmetric combination (called a bonding orbital) is lower in energy than the basis orbitals, and the antisymmetric combination (called an antibonding orbital) is higher. Because the H_2 molecule has two electrons, they can both go in the bonding orbital, making the system lower in energy (and, hence, more stable) than two free hydrogen atoms. This is called a covalent bond. The bond order is equal to the number of bonding electrons minus the number of antibonding electrons, divided by 2. In this example, there are 2 electrons in the bonding orbital and none in the antibonding orbital; the bond order is 1, and there is a single bond between the two hydrogen atoms.

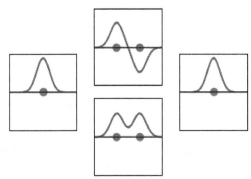

Electron wavefunctions for the 1s orbital of a lone hydrogen atom (left and right) and the corresponding bonding (bottom) and antibonding (top) molecular orbitals of the H_2 molecule. The real part of the wavefunction is the blue curve, and the imaginary part is the red curve. The red dots mark the locations of the nuclei. The electron wavefunction oscillates according to the Schrödinger wave equation, and orbitals are its standing waves. The standing wave frequency is proportional to the orbital's kinetic energy. (This plot is a one-dimensional slice through the three-dimensional system.)

He$_2$

On the other hand, consider the hypothetical molecule of He_2 with the atoms labeled He' and He". As with H_2, the lowest-energy atomic orbitals are the 1s' and 1s", and do not transform according to the symmetries of the molecule, while the symmetry adapted atomic orbitals do. The symmetric combination—the bonding orbital—is lower in energy than the basis orbitals, and the antisymmetric combination—the antibonding orbital—is higher. Unlike H_2, with two valence electrons, He_2 has four in its neutral ground state. Two electrons fill the lower-energy bonding orbital, $\sigma_g(1s)$, while the remaining two fill the higher-energy antibonding orbital, $\sigma_u^*(1s)$. Thus, the resulting electron density around the molecule does not support the formation of a bond between the two atoms; without a stable bond holding the atoms together, molecule would not be expected to exist. Another way of looking at it is that there are two bonding electrons and two antibonding electrons; therefore, the bond order is 0 and no bond exists (the molecule has one bound state supported by the Van der Waals potential).

Li$_2$

Dilithium Li_2 is formed from the overlap of the 1s and 2s atomic orbitals (the basis set) of two Li atoms. Each Li atom contributes three electrons for bonding interactions, and the six electrons fill the three MOs of lowest energy, $\sigma_g(1s)$, $\sigma_u^*(1s)$, and $\sigma_g(2s)$. Using the equation for bond order, it is found that dilithium has a bond order of one, a single bond.

Noble Gases

Considering a hypothetical molecule of He_2, since the basis set of atomic orbitals is the same as in the case of H_2, we find that both the bonding and antibonding orbitals are filled, so there is no energy advantage to the pair. HeH would have a slight energy advantage, but not as much as H_2 + 2 He, so the molecule is very unstable and exists only briefly before decomposing into hydrogen and helium. In general, we find that atoms such as He that have full energy shells rarely bond with other atoms. Except for short-lived Van der Waals complexes, there are very few noble gas compounds known.

Heteronuclear Diatomics

While MOs for homonuclear diatomic molecules contain equal contributions from each interacting atomic orbital, MOs for heteronuclear diatomics contain different atomic orbital contributions. Orbital interactions to produce bonding or antibonding orbitals in heteronuclear diatomics occur if there is sufficient overlap between atomic orbitals as determined by their symmetries and similarity in orbital energies.

HF

In hydrogen fluoride HF overlap between the H 1s and F 2s orbitals is allowed by symmetry but the difference in energy between the two atomic orbitals prevents them from interacting to create a molecular orbital. Overlap between the H 1s and F $2p_z$ orbitals is also symmetry allowed, and these two atomic orbitals have a small energy separation. Thus, they interact, leading to creation of σ and σ^* MOs and a molecule with a bond order of 1. Since HF is a non-centrosymmetric molecule, the symmetry labels g and u do not apply to its molecular orbitals.

Quantitative Approach

To obtain quantitative values for the molecular energy levels, one needs to have molecular orbitals that are such that the configuration interaction (CI) expansion converges fast towards the full CI limit. The most common method to obtain such functions is the Hartree–Fock method, which expresses the molecular orbitals as eigenfunctions of the Fock operator. One usually solves this problem by expanding the molecular orbitals as linear combinations of Gaussian functions centered on the atomic nuclei (linear combination of atomic orbitals and basis set (chemistry)). The equation for the coefficients of these linear combinations is a generalized eigenvalue equation known as the Roothaan equations, which are in fact a particular representation of the Hartree–Fock equation. There are a number of programs in which quantum chemical calculations of MOs can be performed, including Spartan and HyperChem.

Simple accounts often suggest that experimental molecular orbital energies can be obtained by the methods of ultra-violet photoelectron spectroscopy for valence orbitals and X-ray photoelectron spectroscopy for core orbitals. This, however, is incorrect as these experiments measure the ionization energy, the difference in energy between the molecule and one of the ions resulting from the removal of one electron. Ionization energies are linked approximately to orbital energies by Koopmans' theorem. While the agreement between these two values can be close for some molecules, it can be very poor in other cases.

Molecular Orbital Diagram

A molecular orbital diagram, or MO diagram, is a qualitative descriptive tool explaining chemical bonding in molecules in terms of molecular orbital theory in general and the linear combination of atomic orbitals (LCAO) molecular orbital method in particular. A fundamental principle of these theories is that as atoms bond to form molecules, a certain number of atomic orbitals combine to form the same number of molecular orbitals, although the electrons involved may be redistributed among the orbitals. This tool is very well suited for simple diatomic molecules such as dihydrogen, dioxygen, and carbon monoxide but becomes more complex when discussing even comparatively simple polyatomic molecules, such as methane. MO diagrams can explain why some molecules exist and others do not. They can also predict bond strength, as well as the electronic transitions that can take place.

History

Qualitative MO theory was introduced in 1928 by Robert S. Mulliken and Friedrich Hund. A mathematical description was provided by contributions from Douglas Hartree in 1928 and Vladimir Fock in 1930.

Basics

Molecular orbital diagrams are diagrams of molecular orbital (MO) energy levels, shown as short horizontal lines in the center, flanked by constituent atomic orbital (AO) energy levels for comparison, with the energy levels increasing from the bottom to the top. Lines, often dashed diagonal lines, connect MO levels with their constituent AO levels. Degenerate energy levels are commonly shown side by side. Appropriate AO and MO levels are filled with electrons by the Pauli Exclusion Principle, symbolized by small vertical arrows whose directions indicate the electron spins. The AO or MO shapes themselves are often not shown on these diagrams. For a diatomic molecule, an MO diagram effectively shows the energetics of the bond between the two atoms, whose AO unbonded energies are shown on the sides. For simple polyatomic molecules with a "central atom" such as methane (CH_4) or carbon dioxide (CO_2), a MO diagram may show one of the identical bonds to the central atom. For other polyatomic molecules, an MO diagram may show one or more bonds of interest in the molecules, leaving others out for simplicity. Often even for simple molecules, AO and MO levels of inner orbitals and their electrons may be omitted from a diagram for simplicity.

In MO theory molecular orbitals form by the overlap of atomic orbitals. Because σ bonds feature greater overlap than π bonds, σ and σ^* bonding and antibonding orbitals feature greater energy splitting (separation) than π and π^* orbitals. The atomic orbital energy correlates with electronegativity as more electronegative atoms hold their electrons more tightly, lowering their energies. MO modelling is only valid when the atomic ᴜᴧ bltals have comparable energy; when the energies differ greatly the mode of bonding becomes ionic. A second condition for overlapping atomic orbitals is that they have the same symmetry.

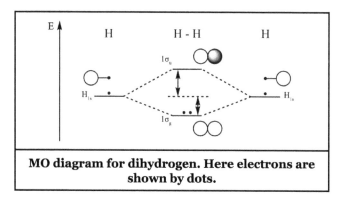

MO diagram for dihydrogen. Here electrons are shown by dots.

Two atomic orbitals can overlap in two ways depending on their phase relationship. The phase of an orbital is a direct consequence of the wave-like properties of electrons. In graphical representations of orbitals, orbital phase is depicted either by a plus or minus sign (which has no relationship to electric charge) or by shading one lobe. The sign of the phase itself does not have physical meaning except when mixing orbitals to form molecular orbitals.

Two same-sign orbitals have a constructive overlap forming a molecular orbital with the bulk of the electron density located between the two nuclei. This MO is called the bonding orbital and its energy is lower than that of the original atomic orbitals. A bond involving molecular orbitals which are symmetric with respect to rotation around the bond axis is called a sigma bond (σ-bond). If the phase changes, the bond becomes a pi bond (π-bond). Symmetry labels are further defined by whether the orbital maintains its original character after an inversion about its center; if it does, it is defined gerade, g. If the orbital does not maintain its original character, it is ungerade, u.

Atomic orbitals can also interact with each other out-of-phase which leads to destructive cancellation and no electron density between the two nuclei at the so-called nodal plane depicted as a perpendicular dashed line. In this anti-bonding MO with energy much higher than the original AO's, any electrons present are located in lobes pointing away from the central internuclear axis. For a corresponding σ-bonding orbital, such an orbital would be symmetrical but differentiated from it by an asterisk as in σ^*. For a π-bond, corresponding bonding and antibonding orbitals would not have such symmetry around the bond axis and be designated π and π^*, respectively.

The next step in constructing an MO diagram is filling the newly formed molecular orbitals with electrons. Three general rules apply:

- The Aufbau principle states that orbitals are filled starting with the lowest energy

- The Pauli exclusion principle states that the maximum number of electrons occupying an orbital is two, with opposite spins

- Hund's rule states that when there are several MO's with equal energy, the electrons occupy the MO's one at a time before two electrons occupy the same MO.

The filled MO highest in energy is called the Highest Occupied Molecular Orbital or HOMO and the empty MO just above it is then the Lowest Unoccupied Molecular Orbital or LUMO. The electrons in the bonding MO's are called bonding electrons and any electrons in the antibonding orbital would be called antibonding electrons. The reduction in energy of these electrons is the driving force for chemical bond formation. Whenever mixing for an atomic orbital is not possible for reasons of symmetry or energy, a non-bonding MO is created, which is often quite similar to and has energy level equal or close to its constituent AO, thus not contributing to bonding energetics. The resulting electron configuration can be described in terms of bond type, parity and occupancy for example dihydrogen $1\sigma_g^2$. Alternatively it can be written as a molecular term symbol e.g. $^1\Sigma_g^+$ for dihydrogen. Sometimes, the letter n is used to designate a non-bonding orbital.

For a stable bond, the bond order, defined as

$$\text{Bond Order} = \frac{(\text{No. of electrons in bonding MOs}) - (\text{No. of electrons in anti-bonding MOs})}{2}$$

must be positive.

The relative order in MO energies and occupancy corresponds with electronic transitions found in photoelectron spectroscopy (PES). In this way it is possible to experimentally verify MO theory. In general, sharp PES transitions indicate nonbonding electrons and broad bands are indicative of bonding and antibonding delocalized electrons. Bands can resolve into fine structure with spacings corresponding to vibrational modes of the molecular cation (see Franck–Condon principle). PES energies are different from ionisation energies which relates to the energy required to strip off the nth electron after the first $n - 1$ electrons have been removed. MO diagrams with energy values can be obtained mathematically using the Hartree–Fock method. The starting point for any MO diagram is a predefined molecular geometry for the molecule in question. An exact relationship between geometry and orbital energies is given in Walsh diagrams.

s-p Mixing

The phenomenon of s-p mixing occurs when molecular orbitals of the same symmetry formed from the combination of 2s and 2p atomic orbitals are close enough in energy to further interact, which can lead to a change in the expected order of orbital energies. When molecular orbitals are formed, they are mathematically obtained from linear combinations of the starting atomic orbitals. Generally, in order to predict their relative energies, it is sufficient to consider only one atomic orbital from each atom to form a pair of molecular orbitals, as the contributions from the others are negligible. For instance, in dioxygen the $3\sigma_g$ MO can be roughly considered to be formed from interaction of oxygen $2p_z$ AOs only. It is found to be lower in energy than the $1\pi_u$ MO, both experimentally and from more sophisticated computational models, so that the expected order of filling is the $3\sigma_g$ before the $1\pi_u$. Hence the approximation to ignore the effects of further interactions is valid. However, experimental and computational results for homonuclear diatomics from Li_2 to N_2 and certain heteronuclear combinations such as CO and NO show that the $3\sigma_g$ MO is higher in energy than (and therefore filled after) the $1\pi_u$ MO. This can be rationalised as the first-approximation $3\sigma_g$ has a suitable symmetry to interact with the $2\sigma_g$ bonding MO formed from the 2s AOs. As a result, the $2\sigma_g$ is lowered in energy, whilst the $3\sigma_g$ is raised. For the aforementioned molecules this results in the $3\sigma_g$ being higher in energy than the $1\pi_u$ MO, which is where s-p mixing is most evident. Likewise, interaction between the $2\sigma_u{}^*$ and $3\sigma_u{}^*$ MOs leads to a lowering in energy of the former and a raising in energy of the latter. However this is of less significance than the interaction of the bonding MOs.

Diatomic MO Diagrams

Dihydrogen

The smallest molecule, hydrogen gas exists as dihydrogen (H-H) with a single covalent bond between two hydrogen atoms. As each hydrogen atom has a single 1s atomic orbital for its electron, the bond forms by overlap of these two atomic orbitals. In figure 1 the two atomic orbitals are depicted on the left and on the right. The vertical axis always represents the orbital energies. Each atomic orbital is singly occupied with an up or down arrow representing an electron.

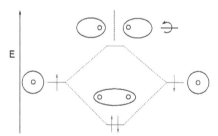

MO diagram of dihydrogen

Application of MO theory for dihydrogen results in having both electrons in the bonding MO with electron configuration $1\sigma_g{}^2$. The bond order for dihydrogen is (2-0)/2 = 1. The photoelectron spectrum of dihydrogen shows a single set of multiplets between 16 and 18 eV (electron volts).

The dihydrogen MO diagram helps explain how a bond breaks. When applying energy to dihydrogen, a molecular electronic transition takes place when one electron in the bonding MO is promoted to the antibonding MO. The result is that there is no longer a net gain in energy.

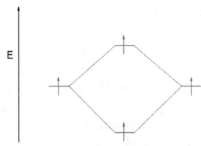

Bond breaking in MO diagram

Dihelium and Diberyllium

Dihelium (He-He) is a hypothetical molecule and MO theory helps to explain why dihelium does not exist in nature. The MO diagram for dihelium looks very similar to that of dihydrogen, but each helium has two electrons in its 1s atomic orbital rather than one for hydrogen, so there are now four electrons to place in the newly formed molecular orbitals.

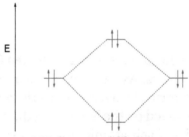

MO diagram of dihelium

The only way to accomplish this is by occupying both the bonding and antibonding orbitals with two electrons, which reduces the bond order $((2-2)/2)$ to zero and cancels the net energy stabilization. However, by removing one electron from dihelium, the stable gas-phase species He_2^+ ion is formed with bond order 1/2.

Another molecule that is precluded based on this principle is diberyllium. Beryllium has an electron configuration $1s^2 2s^2$, so there are again two electrons in the valence level. However, the 2s can mix with the 2p orbitals in diberyllium, whereas there are no p orbitals in the valence level of hydrogen or helium. This mixing makes the antibonding $1\sigma_u$ orbital slightly less antibonding than the bonding $1\sigma_g$ orbital is bonding, with a net effect that the whole configuration has a slight bonding nature. Hence the diberyllium molecule exists (and has been observed in the gas phase). It nevertheless still has a low dissociation energy of only 59 kJ·mol⁻¹.

Dilithium

MO theory correctly predicts that dilithium is a stable molecule with bond order 1 (configuration $1\sigma_g^2 1\sigma_u^2 2\sigma_g^2$). The 1s MOs are completely filled and do not participate in bonding.

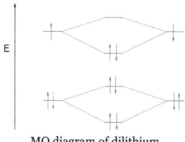

MO diagram of dilithium

Dilithium is a gas-phase molecule with a much lower bond strength than dihydrogen because the 2s electrons are further removed from the nucleus. In a more detailed analysis both the 1σ orbitals have higher energies than the 1s AO and the occupied 2σ is also higher in energy than the 2s AO.

Diboron

The MO diagram for diboron (B-B, electron configuration $1\sigma_g^2 1\sigma_u^2 2\sigma_g^2 2\sigma_u^2 1\pi_u^2$) requires the introduction of an atomic orbital overlap model for p orbitals. The three dumbbell-shaped p-orbitals have equal energy and are oriented mutually perpendicularly (or orthogonally). The p-orbitals oriented in the x-direction (p_x) can overlap end-on forming a bonding (symmetrical) σ orbital and an antibonding σ* molecular orbital. In contrast to the sigma 1s MO's, the σ 2p has some non-bonding electron density at either side of the nuclei and the σ* 2p has some electron density between the nuclei.

The other two p-orbitals, p_y and p_z, can overlap side-on. The resulting bonding orbital has its electron density in the shape of two lobes above and below the plane of the molecule. The orbital is not symmetric around the molecular axis and is therefore a pi orbital. The antibonding pi orbital (also asymmetrical) has four lobes pointing away from the nuclei. Both p_y and p_z orbitals form a pair of pi orbitals equal in energy (degenerate) and can have higher or lower energies than that of the sigma orbital.

In diboron the 1s and 2s electrons do not participate in bonding but the single electrons in the 2p orbitals occupy the $2\pi p_y$ and the $2\pi p_z$ MO's resulting in bond order 1. Because the electrons have equal energy (they are degenerate) diboron is a diradical and since the spins are parallel the compound is paramagnetic.

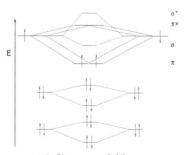

MO diagram of diboron

In certain diborynes the boron atoms are excited and the bond order is 3.

Dicarbon

Like diboron, dicarbon (C-C electron configuration:$1\sigma_g^2 1\sigma_u^2 2\sigma_g^2 2\sigma_u^2 1\pi_u^4$) is a reactive gas-phase molecule. The molecule can be described as having two pi bonds but without a sigma bond.

Dinitrogen

The bond order for dinitrogen ($1\sigma_g^2 1\sigma_u^2 2\sigma_g^2 2\sigma_u^2 1\pi_u^4 3\sigma_g^2$) is three because two electrons are now also added in the 3σ MO. The MO diagram correlates with the experimental photoelectron spectrum for nitrogen. The 1σ electrons can be matched to a peak at 410 eV (broad), the $2\sigma_g$ electrons at 37 eV (broad), the $2\sigma_u$ electrons at 19 eV (doublet), the $1\pi_u^4$ electrons at 17 eV (multiplets), and finally the $3\sigma_g^2$ at 15.5 eV (sharp).

Dioxygen

MO treatment of dioxygen is different from that of the previous diatomic molecules because the $p\sigma$ MO is now lower in energy than the 2π orbitals. This is attributed to interaction between the 2s MO and the $2p_z$ MO. Distributing 8 electrons over 6 molecular orbitals leaves the final two electrons as a degenerate pair in the $2p\pi^*$ antibonding orbitals resulting in a bond order of 2. As in diboron, these two unpaired electrons have the same spin in the ground state, which is a paramagnetic diradical triplet oxygen. The first excited state has both HOMO electrons paired in one orbital with opposite spins, and is known as singlet oxygen.

The bond order decreases and the bond length increases in the order O_2^+ (112.2 pm), O_2 (121 pm), O_2^- (128 pm) and O_2^{2-} (149 pm).

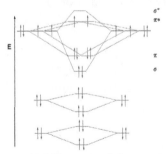

MO diagram of dioxygen triplet ground state

Difluorine and Dineon

In difluorine two additional electrons occupy the $2p\pi^*$ with a bond order of 1. In dineon Ne_2 (as with dihelium) the number of bonding electrons equals the number of antibonding electrons and this compound does not exist.

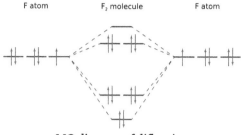

MO diagram of difluorine

Dimolybdenum and Ditungsten

Dimolybdenum (Mo_2) is notable for having a sextuple bond. This involves two sigma bonds ($4d_z^2$ and $5s$), two pi bonds (using $4d_{xz}$ and $4d_{yz}$), and two delta bonds ($4d_{x^2-y^2}$ and $4d_{xy}$). Ditungsten (W_2) has a similar structure.

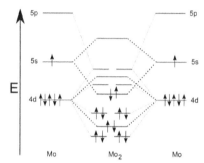

MO diagram of dimolybdenum

Heteronuclear Diatomics

In heteronuclear diatomic molecules, mixing of atomic orbitals only occurs when the electronegativity values are similar. In carbon monoxide (CO, isoelectronic with dinitrogen) the oxygen 2s orbital is much lower in energy than the carbon 2s orbital and therefore the degree of mixing is low. The electron configuration $1\sigma^2 1\sigma^{*2} 2\sigma^2 2\sigma^{*2} 1\pi^4 3\sigma^2$ is identical to that of nitrogen. The g and u subscripts no longer apply because the molecule lacks a center of symmetry.

In hydrogen fluoride (HF), the hydrogen 1s orbital can mix with fluorine $2p_z$ orbital to form a sigma bond because experimentally the energy of 1s of hydrogen is comparable with 2p of fluorine. The HF electron configuration $1\sigma^2 2\sigma^2 3\sigma^2 1\pi^4$ reflects that the other electrons remain in three lone pairs and that the bond order is 1.

Triatomic Molecules

Carbon Dioxide

Carbon dioxide, CO_2, is a linear molecule with a total of sixteen bonding electrons in its valence shell. Carbon is the central atom of the molecule and a principal axis, the z-axis, is visualized as a single axis that goes through the center of carbon and the two oxygens atoms. For convention, blue atomic orbital lobes are positive phases, red atomic orbitals are negative phases, with respect to the wave function from the solution of the Schrödinger equation. In carbon dioxide the carbon 2s (−19.4 eV), carbon 2p (−10.7 eV), and oxygen 2p (−15.9 eV)) energies associated with the atomic orbitals are in proximity whereas the oxygen 2s energy (−32.4 eV) is different.

Carbon and each oxygen atom will have a 2s atomic orbital and a 2p atomic orbital, where the p orbital is divided into p_x, p_y, and p_z. With these derived atomic orbitals, symmetry labels are deduced with respect to rotation about the principal axis which generates a phase change, pi bond (π) or generates no phase change, known as a sigma bond (σ). Symmetry labels are further defined by whether the atomic orbital maintains its original character after an inversion about its center atom; if the atomic orbital does retain its original character it is defined gerade, g, or if the atomic orbital does not maintain its original character, ungerade, u. The final symmetry-labeled atomic orbital is now known as an irreducible representation.

Carbon dioxide's molecular orbitals are made by the linear combination of atomic orbitals of the same irreducible representation that are also similar in atomic orbital energy. Significant atomic orbital overlap explains why sp bonding may occur. Strong mixing of the oxygen 2s atomic orbital is not to be expected and are non-bonding degenerate molecular orbitals. The combination of similar atomic orbital/wave functions and the combinations of atomic orbital/wave function inverses create particular energies associated with the nonbonding (no change), bonding (lower than either parent orbital energy) and antibonding (higher energy than either parent atomic orbital energy) molecular orbitals.

MO model carbon dioxide

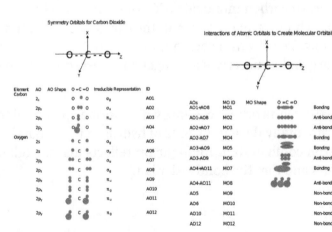
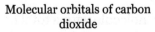
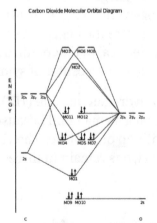

| Atomic orbitals of carbon dioxide | Molecular orbitals of carbon dioxide | MO Diagram of carbon dioxide |

Water

Water (H_2O) is a bent molecule ($105°$) with C_{2v} molecular symmetry. The oxygen atomic orbitals are labeled according to their symmetry as a_1 for the 2s orbital and b_1 ($2p_x$), b_2 ($2p_y$) and a_1 ($2p_z$) for the three 2p orbitals. The two hydrogen 1s orbitals are premixed to form a_1 (σ) and b_2 (σ^*) MO.

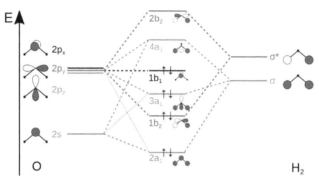

Molecular orbital diagram of water

C_{2v}	E	C_2	$\sigma_v(xz)$	$\sigma_v'(yz)$		
A_1	1	1	1	1	z	x^2, y^2, z^2
A_2	1	1	−1	−1	R_z	xy
B_1	1	−1	1	−1	x, R_y	xz
B_2	1	−1	−1	1	y, R_x	yz

Mixing takes place between same-symmetry orbitals of comparable energy resulting a new set of MO's for water:

- $2a_1$ MO from mixing of the oxygen 2s AO and the hydrogen σ MO. Small oxygen $2p_z$ AO admixture strengthens bonding and lowers the orbital energy.

- $1b_2$ MO from mixing of the oxygen $2p_y$ AO and the hydrogen σ^* MO.

- $3a_1$ MO from mixing of the oxygen $2p_z$ AO and the hydrogen σ MO. Small oxygen 2s AO admixture weakens bonding and raises the orbital energy.

- $1b_1$ nonbonding MO from the oxygen $2p_x$ AO (the p-orbital perpendicular to the molecular plane).

In agreement with this description the photoelectron spectrum for water shows a sharp peak for the nonbonding $1b_1$ MO (12.6 eV) and three broad peaks for the $3a_1$ MO (14.7 eV), $1b_2$ MO (18.5 eV) and the $2a_1$ MO (32.2 eV). The $1b_1$ MO is a lone pair, while the $3a_1$, $1b_2$ and $2a_1$ MO's can be localized to give two O–H bonds and an in-plane lone pair. This MO treatment of water does not have two equivalent *rabbit ear* lone pairs.

Hydrogen sulfide (H_2S) too has a C_{2v} symmetry with 8 valence electrons but the bending angle is only 92°. As reflected in its photoelectron spectrum as compared to water the $5a_1$ MO (corresponding to the $3a_1$ MO in water) is stabilised (improved overlap) and the $2b_2$ MO (corresponding to the $1b_2$ MO in water) is destabilized (poorer overlap).

Hydrogen Atom

A hydrogen atom is an atom of the chemical element hydrogen. The electrically neutral atom contains a single positively charged proton and a single negatively charged electron bound to the nucleus by the Coulomb force. Atomic hydrogen constitutes about 75% of the elemental (baryonic) mass of the universe.

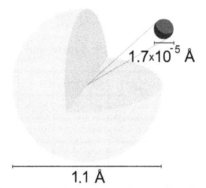

1.7×10^{-5} Å

1.1 Å

Depiction of a hydrogen atom showing the diameter as about twice the Bohr model radius. (Image not to scale)

In everyday life on Earth, isolated hydrogen atoms (usually called "atomic hydrogen" or, more precisely, "monatomic hydrogen") are extremely rare. Instead, hydrogen tends to combine with other atoms in compounds, or with itself to form ordinary (diatomic) hydrogen gas, H_2. "Atomic hydrogen" and "hydrogen atom" in ordinary English use have overlapping, yet distinct, meanings. For example, a water molecule contains two hydrogen atoms, but does not contain atomic hydrogen (which would refer to isolated hydrogen atoms).

Attempts to develop a theoretical understanding of the hydrogen atom have been important to the history of quantum mechanics.

Isotopes

The most abundant isotope, hydrogen-1, protium, or light hydrogen, contains no neutrons and is just a proton and an electron. Protium is stable and makes up 99.9885% of naturally occurring hydrogen by absolute number (not mass).

Deuterium contains one neutron and one proton. Deuterium is stable and makes up

0.0115% of naturally occurring hydrogen and is used in industrial processes like nuclear reactors and Nuclear Magnetic Resonance.

Tritium contains two neutrons and one proton and is not stable, decaying with a half-life of 12.32 years. Because of the short half life, Tritium does not exist in nature except in trace amounts.

Higher isotopes of hydrogen are only created in artificial accelerators and reactors and have half lives around the order of 10^{-22} seconds.

The formulas below are valid for all three isotopes of hydrogen, but slightly different values of the Rydberg constant (correction formula given below) must be used for each hydrogen isotope.

Hydrogen Ion

Hydrogen is not found without its electron in ordinary chemistry (room temperatures and pressures), as ionized hydrogen is highly chemically reactive. When ionized hydrogen is written as "H^+" as in the solvation of classical acids such as hydrochloric acid, the hydronium ion, H_3O^+, is meant, not a literal ionized single hydrogen atom. In that case, the acid transfers the proton to H_2O to form H_3O^+.

Ionized hydrogen without its electron, or free protons, are common in the interstellar medium, and solar wind.

Theoretical Analysis

The hydrogen atom has special significance in quantum mechanics and quantum field theory as a simple two-body problem physical system which has yielded many simple analytical solutions in closed-form.

Failed Classical Description

Experiments by Ernest Rutherford in 1909 showed the structure of the atom to be a dense, positive nucleus with a light, negative charge orbiting around it. This immediately caused problems on how such a system could be stable. Classical electromagnetism had shown that any accelerating charge radiates energy described through the Larmor formula. If the electron is assumed to orbit in a perfect circle and radiates energy continuously, the electron would rapidly spiral into the nucleus with a fall time of:

$$t_{\text{fall}} \approx \frac{a_0^3}{4r_0^2 c} \approx 1.6 \cdot 10^{-11} \text{s}$$

Where a_0 is the Bohr radius and r_0 is the classical electron radius. If this were true, all atoms would instantly collapse, however atoms seem to be stable. Furthermore, the spiral inward would release a smear of electromagnetic frequencies as the orbit got smaller. Instead, atoms were observed to only emit discrete frequencies of light. The resolution would lie in the development of quantum mechanics.

Bohr-Sommerfeld Model

In 1913, Niels Bohr obtained the energy levels and spectral frequencies of the hydrogen atom after making a number of simple assumptions in order to correct the failed classical model. The assumptions included:

1. Electrons can only be in certain, discrete circular orbits or *stationary states*, thereby having a discrete set of possible radii and energies.

2. Electrons do not emit radiation while in one of these stationary states.

3. An electron can gain or lose energy by jumping from one discrete orbital to another.

Bohr supposed that the electron's angular momentum is quantized with possible values:

$L = n\hbar$ where $n = 1, 2, 3, \ldots$

and $\hbar$ is Planck constant over 2π. He also supposed that the centripetal force which keeps the electron in its orbit is provided by the Coulomb force, and that energy is conserved. Bohr derived the energy of each orbit of the hydrogen atom to be:

$$E_n = -\frac{m_e e^4}{2(4\pi\varepsilon_0)^2 \hbar^2} \frac{1}{n^2},$$

where m_e is the electron mass, e is the electron charge, ε_0 is the electric permeability, and n is the quantum number (now known as the principal quantum number). Bohr's predictions matched experiments measuring the hydrogen spectral series to the first order, giving more confidence to a theory that used quantized values.

For $n = 1$, the value

$$\frac{m_e e^4}{2(4\pi\epsilon_0)^2 \hbar^2} = \frac{m_e e^4}{8h^2\varepsilon_0^2} = 1Ry = 13.605\,692\,53(30)\text{eV}$$

is called the Rydberg unit of energy. It is related to the Rydberg constant R_∞ of atomic physics by $1Ry \equiv hcR_\infty$.

The exact value of the Rydberg constant assumes that the nucleus is infinitely massive with respect to the electron. For hydrogen-1, hydrogen-2 (deuterium), and hydrogen-3 (tritium) the constant must be slightly modified to use the reduced mass of the system, rather than simply the mass of the electron. However, since the nucleus is much heavier than the electron, the values are nearly the same. The Rydberg constant R_M for a hydrogen atom (one electron), R is given by

$$R_M = \frac{R_\infty}{1 + m_e / M},$$

where M is the mass of the atomic nucleus. For hydrogen-1, the quantity $m_e >$ is about 1/1836 (i.e. the electron-to-proton mass ratio). For deuterium and tritium, the ratios are about 1/3670 and 1/5497 respectively. These figures, when added to 1 in the denominator, represent very small corrections in the value of R, and thus only small corrections to all energy levels in corresponding hydrogen isotopes.

There were still problems with Bohr's model:

1. it failed to predict other spectral details such as fine structure and hyperfine structure

2. it could only predict energy levels with any accuracy for single–electron atoms (hydrogen–like atoms)

3. the predicted values were only correct to $\alpha^2 \approx 10^{-5}$, where α is the fine-structure constant.

Most of these shortcomings were repaired by Arnold Sommerfeld's modification of the Bohr model. Sommerfeld introduced two additional degrees of freedom allowing an electron to move on an elliptical orbit, characterized by its eccentricity and declination with respect to a chosen axis. This introduces two additional quantum numbers, which correspond to the orbital angular momentum and its projection on the chosen axis. Thus the correct multiplicity of states (except for the factor 2 accounting for the yet unknown electron spin) was found. Further applying special relativity theory to the elliptic orbits, Sommerfeld succeeded in deriving the correct expression for the fine structure of hydrogen spectra (which happens to be exactly the same as in the most elaborate Dirac theory). However some observed phenomena such as the anomalous Zeeman effect remain unexplained. These issues were resolved with the full development of quantum mechanics and the Dirac equation. It is often alleged, that the Schrödinger equation is superior to the Bohr-Sommerfeld theory in describing hydrogen atom. This is however not the case, as the most results of both approaches coincide or are very close (a remarkable exception is the problem of hydrogen atom in crossed electric and magnetic fields, which cannot be solved in the framework of the Bohr-Sommerfeld theory self-consistently), and their main shortcomings result from the absence of the electron spin in both theories. It was the complete failure of the Bohr-Sommerfeld theory to explain many-electron systems (such as helium atom or hydrogen molecule) which demonstrated its inadequacy in describing quantum phenomena.

Schrödinger Equation

The Schrödinger equation allows one to calculate the development of quantum systems with time and can give exact, analytical answers for the non-relativistic hydrogen atom.

Wavefunction

The Hamiltonian of the hydrogen atom is the radial kinetic energy operator and coulomb attraction force between the positive proton and negative electron. Using the time-independent Schrödinger equation, ignoring all spin-coupling interactions and using the reduced mass μ, the equation is written as:

$$\left(-\frac{\hbar^2}{2\mu}\nabla^2 - \frac{Ze^2}{4\pi\varepsilon_0 r}\right)\psi(r,\theta,\phi) = E\psi(r,\theta,\phi)$$

Expanding the Laplacian in spherical coordinates:

$$-\frac{\hbar^2}{2\mu}\left[\frac{1}{r^2}\frac{\partial}{\partial r}\left(r^2\frac{\partial\psi}{\partial r}\right) + \frac{1}{r^2\sin\theta}\frac{\partial}{\partial\theta}\left(\sin\theta\frac{\partial\psi}{\partial\theta}\right) + \frac{1}{r^2\sin^2\theta}\frac{\partial^2\psi}{\partial\phi^2}\right] - \frac{Ze^2}{4\pi\varepsilon_0 r}\psi = E\psi$$

This is a separable, partial differential equation which can be solved in terms of special functions. By setting Z=1 (for one proton), the normalized position wavefunctions, given in spherical coordinates are:

$$\psi_{n\ell m}(r,\vartheta,\varphi) = \sqrt{\left(\frac{2}{na_0}\right)^3\frac{(n-\ell-1)!}{2n[(n+\ell)!]}}e^{-\rho/2}\rho^\ell L_{n-\ell-1}^{2\ell+1}(\rho)Y_\ell^m(\vartheta,\varphi)$$

3D illustration of the eigenstate $\psi_{4,3,1}$. Electrons in this state are 45% likely to be found within the solid body shown.

where:

$$\rho = \frac{2r}{na_0},$$

a_0 is the Bohr radius,

$L_{n-\ell-1}^{2\ell+1}(\rho)$ is a generalized Laguerre polynomial of degree $n-\ell-1$, and

$Y_\ell^m(\vartheta,\varphi)$ is a spherical harmonic function of degree ℓ and order m. Note that the generalized Laguerre polynomials are defined differently by different authors. The usage here is consistent with the definitions used by Messiah, and Mathematica. In other places, the Laguerre polynomial includes a factor of $(n+\ell)!$, or the generalized Laguerre polynomial appearing in the hydrogen wave function is $L_{n+\ell}^{2\ell+1}(\rho)$ instead.

The quantum numbers can take the following values:

$$n = 1, 2, 3, \ldots$$

$$\ell = 0, 1, 2, \ldots, n-1$$

$$m = -\ell, \ldots, \ell.$$

Additionally, these wavefunctions are *normalized* (i.e., the integral of their modulus square equals 1) and orthogonal:

$$\int_0^\infty r^2 dr \int_0^\pi \sin\vartheta d\vartheta \int_0^{2\pi} d\varphi \, \psi_{n\ell m}^*(r,\vartheta,\varphi)\psi_{n'\ell'm'}(r,\vartheta,\varphi) = \langle n, \ell, m \mid n', \ell', m' \rangle = \delta_{nn'}\delta_{\ell\ell'}\delta_{mm'},$$

where $\mid n, \ell, m \rangle$ is the state represented by the wavefunction $\psi_{n\ell m}$ in Dirac notation, and δ is the Kronecker delta function.

The wavefunctions in momentum space are related to the wavefunctions in position space through a Fourier transform

$$\phi(p,\vartheta_p,\varphi_p) = (2\pi\hbar)^{-3/2}\int e^{-i\vec{p}\cdot\vec{r}/\hbar}\psi(r,\vartheta,\varphi)dV,$$

which, for the bound states, results in

$$\phi(p,\vartheta_p,\varphi_p) = \sqrt{\frac{2}{\pi}\frac{(n-l-1)!}{(n+l)!}}n^2 2^{2l+2}l!\frac{n^l p^l}{(n^2 p^2 +1)^{l+2}}C_{n-l-1}^{l+1}\left(\frac{n^2 p^2 -1}{n^2 p^2 +1}\right)Y_l^m(\vartheta_p,\varphi_p),$$

where $C_N^\alpha(x)$ denotes a Gegenbauer polynomial and p is in units of $\hbar/a_0$.

The solutions to the Schrödinger equation for hydrogen are analytical, giving a simple expression for the hydrogen energy levels and thus the frequencies of the hydrogen spectral lines and fully reproduced the Bohr model and went beyond it. It also yields two other quantum numbers and the shape of the electron's wave function ("orbital") for the various possible quantum-mechanical states, thus explaining the anisotropic character of atomic bonds.

The Schrödinger equation also applies to more complicated atoms and molecules. When there is more than one electron or nucleus the solution is not analytical and either computer calculations are necessary or simplifying assumptions must be made.

Since the Schrödinger equation is only valid for non-relativistic quantum mechanics, the solutions it yields for the hydrogen atom are not entirely correct. The Dirac equation of relativistic quantum theory improves these solutions.

Results of Schrödinger Equation

The solution of the Schrödinger equation (wave equation) for the hydrogen atom uses the fact that the Coulomb potential produced by the nucleus is isotropic (it is radially symmetric in space and only depends on the distance to the nucleus). Although the resulting energy eigenfunctions (the *orbitals*) are not necessarily isotropic themselves, their dependence on the angular coordinates follows completely generally from this isotropy of the underlying potential: the eigenstates of the Hamiltonian (that is, the energy eigenstates) can be chosen as simultaneous eigenstates of the angular momentum operator. This corresponds to the fact that angular momentum is conserved in the orbital motion of the electron around the nucleus. Therefore, the energy eigenstates may be classified by two angular momentum quantum numbers, ℓ and m (both are integers). The angular momentum quantum number $\ell = 0, 1, 2, ...$ determines the magnitude of the angular momentum. The magnetic quantum number $m = -\ell, ..., +\ell$ determines the projection of the angular momentum on the (arbitrarily chosen) z-axis.

In addition to mathematical expressions for total angular momentum and angular momentum projection of wavefunctions, an expression for the radial dependence of the wave functions must be found. It is only here that the details of the $1/r$ Coulomb potential enter (leading to Laguerre polynomials in r). This leads to a third quantum number, the principal quantum number $n = 1, 2, 3, ...$ The principal quantum number in hydrogen is related to the atom's total energy.

Note that the maximum value of the angular momentum quantum number is limited by the principal quantum number: it can run only up to $n - 1$, i.e. $\ell = 0, 1, ..., n - 1$.

Due to angular momentum conservation, states of the same ℓ but different m have the same energy (this holds for all problems with rotational symmetry). In addition, for the hydrogen atom, states of the same n but different ℓ are also degenerate (i.e. they have the same energy). However, this is a specific property of hydrogen and is no longer true for more complicated atoms which have an (effective) potential differing from the form $1/r$ (due to the presence of the inner electrons shielding the nucleus potential).

Taking into account the spin of the electron adds a last quantum number, the projection of the electron's spin angular momentum along the z-axis, which can take on two values. Therefore, any eigenstate of the electron in the hydrogen atom is described fully by four quantum numbers. According to the usual rules of quantum mechanics, the actual state of the electron may be any superposition of these states. This explains also why the choice of z-axis for the directional quantization of the angular momentum vector is immaterial: an orbital of given ℓ and m' obtained for another preferred axis z' can always be represented as a suitable superposition of the various states of different m (but same l) that have been obtained for z.

Mathematical Summary of Eigenstates of Hydrogen Atom

In 1928, Paul Dirac found an equation that was fully compatible with Special Relativity, and (as a consequence) made the wave function a 4-component "Dirac spinor" including "up" and "down" spin components, with both positive and "negative" energy (or matter and antimatter). The solution to this equation gave the following results, more accurate than the Schrödinger solution.

Energy Levels

The energy levels of hydrogen, including fine structure (excluding Lamb shift and hyperfine structure), are given by the Sommerfeld fine structure expression:

where α is the fine-structure constant and j is the "total angular momentum" quantum number, which is equal to $|\ell \pm \frac{1}{2}|$ depending on the direction of the electron spin. This formula represents a small correction to the energy obtained by Bohr and Schrödinger as given above. The factor in square brackets in the last expression is nearly one; the extra term arises from relativistic effects. It is worth noting that this expression was first obtained by A. Sommerfeld in 1916 based on the relativistic version of the old Bohr theory. Sommer-feld has however used different notation for the quantum numbers.

$$E_{jn} = -m_e c^2 \left[1 - \left(1 + \left[\frac{\alpha}{n - j - \frac{1}{2} + \sqrt{\left(j + \frac{1}{2}\right)^2 - \alpha^2}} \right]^2 \right)^{-1/2} \right]$$

$$\approx -\frac{m_e c^2 \alpha^2}{2n^2} \left[1 + \frac{\alpha^2}{n^2} \left(\frac{n}{j + \frac{1}{2}} - \frac{3}{4} \right) \right],$$

Visualizing the Hydrogen Electron Orbitals

The image to the right shows the first few hydrogen atom orbitals (energy eigenfunctions). These are cross-sections of the probability density that are color-coded (black represents zero density and white represents the highest density). The angular momentum (orbital) quantum number ℓ is denoted in each column, using the usual spectroscopic letter code (s means $\ell = 0$, p means $\ell = 1$, d means $\ell = 2$). The main (principal) quantum number n (= 1, 2, 3, ...) is marked to the right of each row. For all pictures the magnetic quantum number m has been set to 0, and the cross-sectional plane is the xz-plane (z is the vertical axis). The probability density in three-dimensional space is obtained by rotating the one shown here around the z-axis.

The "ground state", i.e. the state of lowest energy, in which the electron is usually found, is the first one, the 1s state (principal quantum level $n = 1$, $\ell = 0$).

Black lines occur in each but the first orbital: these are the nodes of the wavefunction, i.e. where the probability density is zero. (More precisely, the nodes are spherical harmonics that appear as a result of solving Schrödinger's equation in polar coordinates.)

The quantum numbers determine the layout of these nodes. There are:

- $n-1$ total nodes,

- l of which are angular nodes:

  - m angular nodes go around the ϕ axis (in the xy plane). (The figure above does not show these nodes since it plots cross-sections through the xz-plane.)

  - $l-m$ (the remaining angular nodes) occur on the θ (vertical) axis.

- $n-l-1$ (the remaining non-angular nodes) are radial nodes.

Features Going Beyond the Schrödinger Solution

There are several important effects that are neglected by the Schrödinger equation and which are responsible for certain small but measurable deviations of the real spectral lines from the predicted ones:

- Although the mean speed of the electron in hydrogen is only 1/137th of the speed of light, many modern experiments are sufficiently precise that a complete theoretical explanation requires a fully relativistic treatment of the problem. A relativistic treatment results in a momentum increase of about 1 part in 37,000 for the electron. Since the electron's wavelength is determined by its momentum, orbitals containing higher speed electrons show contraction due to smaller wavelengths.

- Even when there is no external magnetic field, in the inertial frame of the moving electron, the electromagnetic field of the nucleus has a magnetic component. The spin of the electron has an associated magnetic moment which interacts with this magnetic field. This effect is also explained by special relativity, and it leads to the so-called *spin-orbit coupling*, i.e., an interaction between the electron's orbital motion around the nucleus, and its spin.

Both of these features (and more) are incorporated in the relativistic Dirac equation, with predictions that come still closer to experiment. Again the Dirac equation may be solved analytically in the special case of a two-body system, such as the hydrogen atom. The resulting solution quantum states now must be classified by the total angular momentum number j (arising through the coupling between electron spin and orbital angular momentum). States of the same j and the same n are still degenerate. Thus, direct analytical solution of Dirac equation predicts $2S(\frac{1}{2})$ and $2P(\frac{1}{2})$ levels of Hydrogen to have exactly the same energy, which is in a contradiction with observations (Lamb-Retherford experiment).

- There are always vacuum fluctuations of the electromagnetic field, according to quantum mechanics. Due to such fluctuations degeneracy between states of the same j but different l is lifted, giving them slightly different energies. This has been demonstrated in the famous Lamb-Retherford experiment and was the starting point for the development of the theory of Quantum electrodynamics (which is able to deal with these vacuum fluctuations and employs the famous Feynman diagrams for approximations using perturbation theory). This effect is now called Lamb shift.

For these developments, it was essential that the solution of the Dirac equation for the hydrogen atom could be worked out exactly, such that any experimentally observed deviation had to be taken seriously as a signal of failure of the theory.

Alternatives to the Schrödinger Theory

In the language of Heisenberg's matrix mechanics, the hydrogen atom was first solved by Wolfgang Pauli using a rotational symmetry in four dimension [O(4)-symmetry] generated by the angular momentum and the Laplace–Runge–Lenz vector. By extending the symmetry group O(4) to the dynamical group O(4,2), the entire spectrum and all transitions were embedded in a single irreducible group representation.

In 1979 the (non relativistic) hydrogen atom was solved for the first time within Feynman's path integral formulation of quantum mechanics. This work greatly extended the range of applicability of Feynman's method.

Hydrogen-like Atom

A hydrogen-like ion is any atomic nucleus with one electron and thus is isoelectronic with hydrogen. Except for the hydrogen atom itself (which is neutral), these ions carry the positive charge $e(Z-1)$, where Z is the atomic number of the atom. Examples of hydrogen-like ions are He^+, Li^{2+}, Be^{3+} and B^{4+}. Because hydrogen-like ions are two-particle systems with an interaction depending only on the distance between the two particles, their (non-relativistic) Schrödinger equation can be solved in analytic form, as can the (relativistic) Dirac equation. The solutions are one-electron functions and are referred to as *hydrogen-like atomic orbitals*.

Other systems may also be referred to as "hydrogen-like atoms", such as muonium (an electron orbiting a muon), positronium (an electron and a positron), certain exotic atoms (formed with other particles), or Rydberg atoms (in which one electron is in such a high energy state that it sees the rest of the atom practically as a point charge).

Schrödinger Solution

In the solution to the Schrödinger equation, which is non-relativistic, hydrogen-like atomic orbitals are eigenfunctions of the one-electron angular momentum operator L and its z component L_z. A hydrogen-like atomic orbital is uniquely identified by the values of the principal quantum number n, the angular momentum quantum number l, and the magnetic quantum number m. The energy eigenvalues do not depend on l or m, but solely on n. To these must be added the two-valued spin quantum number m_s = ±½, setting the stage for the Aufbau principle. This principle restricts the allowed values of the four quantum numbers in electron configurations of more-electron atoms. In hydrogen-like atoms all degenerate orbitals of fixed n and l, m and s varying between certain values form an atomic shell.

The Schrödinger equation of atoms or atomic ions with more than one electron has not been solved analytically, because of the computational difficulty imposed by the Coulomb interaction between the electrons. Numerical methods must be applied in order to obtain (approximate) wavefunctions or other properties from quantum mechanical calculations. Due to the spherical symmetry (of the Hamiltonian), the total angular momentum J of an atom is a conserved quantity. Many numerical procedures start from products of atomic orbitals that are eigenfunctions of the one-electron operators L and L_z. The radial parts of these atomic orbitals are sometimes numerical tables or are sometimes Slater orbitals. By angular momentum coupling many-electron eigenfunctions of J^2 (and possibly S^2) are constructed.

In quantum chemical calculations hydrogen-like atomic orbitals cannot serve as an expansion basis, because they are not complete. The non-square-integrable continuum ($E > 0$) states must be included to obtain a complete set, i.e., to span all of one-electron Hilbert space.

In the simplest model, the atomic orbitals of hydrogen-like ions are solutions to the Schrödinger equation in a spherically symmetric potential. In this case, the potential term is the potential given by Coulomb's law:

$$V(r) = -\frac{1}{4\pi\varepsilon_0}\frac{Ze^2}{r}$$

where

- ε_0 is the permittivity of the vacuum,

- Z is the atomic number (number of protons in the nucleus),

- e is the elementary charge (charge of an electron),

- r is the distance of the electron from the nucleus.

After writing the wave function as a product of functions:

$$\psi(r,\theta,\phi) = R_{nl}(r)Y_{lm}(\theta,\phi)$$

(in spherical coordinates), where Y_{lm} are spherical harmonics, we arrive at the following Schrödinger equation:

$$\left[-\frac{\hbar^2}{2\mu} \left(\frac{1}{r^2} \frac{\partial}{\partial r} \left(r^2 \frac{\partial R(r)}{\partial r} \right) - \frac{l(l+1)R(r)}{r^2} \right) + V(r)R(r) \right] = ER(r),$$

where μ is, approximately, the mass of the electron (more accurately, it is the reduced mass of the system consisting of the electron and the nucleus), and $\hbar$ is the reduced Planck constant.

Different values of l give solutions with different angular momentum, where l (a non-negative integer) is the quantum number of the orbital angular momentum. The magnetic quantum number m (satisfying $-l \le m \le l$) is the (quantized) projection of the orbital angular momentum on the z-axis.

Non-relativistic Wavefunction and Energy

In addition to l and m, a third integer $n > 0$, emerges from the boundary conditions placed on R. The functions R and Y that solve the equations above depend on the values of these integers, called *quantum numbers*. It is customary to subscript the wave functions with the values of the quantum numbers they depend on. The final expression for the normalized wave function is:

$$\psi_{nlm} = R_{nl}(r)Y_{lm}(\theta, \phi)$$

$$R_{nl}(r) = \sqrt{\left(\frac{2Z}{na_\mu} \right)^3 \frac{(n-l-1)!}{2n[(n+l)!]}} e^{-Zr/na_\mu} \left(\frac{2Zr}{na_\mu} \right)^l L_{n-l-1}^{2l+1}\left(\frac{2Zr}{na_\mu} \right)$$

where:

- L_{n-l-1}^{2l+1} are the generalized Laguerre polynomials in the definition given here.

- $a_\mu = \frac{4\pi\varepsilon_0 \hbar^2}{\mu e^2} = \frac{\hbar c}{\alpha\mu c^2} = \frac{m_e}{\mu} a_0$

  where α is the fine structure constant. Here, μ is the reduced mass of the nucleus-electron system, that is, $\mu = \frac{m_N m_e}{m_N + m_e}$ where m_N is the mass of the nucleus. Typically, the nucleus is much more massive than the electron, so $\mu \approx m_e$. (But for positronium $\mu = m_e / 2$.)

- $E_n = -\left(\frac{Z^2 \mu e^4}{32\pi^2 \varepsilon_0^2 \hbar^2} \right) \frac{1}{n^2} = -\left(\frac{Z^2 \hbar^2}{2\mu a_\mu^2} \right) \frac{1}{n^2} = -\frac{\mu c^2 Z^2 \alpha^2}{2n^2}.$

- $Y_{lm}(\theta, \phi)$ function is a spherical harmonic.

parity due to angular wave function is $(-1)^l$.

Quantum Numbers

The quantum numbers n, l and m are integers and can have the following values:

$$n = 1, 2, 3, 4, \ldots$$

$$l = 0, 1, 2, \ldots, n-1$$

$$m = -l, -l+1, \ldots, 0, \ldots, l-1, l$$

For a group-theoretical interpretation of these quantum numbers. Among other things, this article gives group-theoretical reasons why $l < n$ and $-l \leq m \leq l$.

Angular Momentum

Each atomic orbital is associated with an angular momentum **L**. It is a vector operator, and the eigenvalues of its square $L^2 \equiv L_x{}^2 + L_y{}^2 + L_z{}^2$ are given by:

$$L^2 Y_{lm} = \hbar^2 l(l+1) Y_{lm}$$

The projection of this vector onto an arbitrary direction is quantized. If the arbitrary direction is called z, the quantization is given by:

$$L_z Y_{lm} = \hbar m Y_{lm},$$

where m is restricted as described above. Note that L^2 and L_z commute and have a common eigenstate, which is in accordance with Heisenberg's uncertainty principle. Since L_x and L_y do not commute with L_z, it is not possible to find a state that is an eigenstate of all three components simultaneously. Hence the values of the x and y components are not sharp, but are given by a probability function of finite width. The fact that the x and y components are not well-determined, implies that the direction of the angular momentum vector is not well determined either, although its component along the z-axis is sharp.

These relations do not give the total angular momentum of the electron. For that, electron spin must be included.

This quantization of angular momentum closely parallels that proposed by Niels Bohr (see Bohr model) in 1913, with no knowledge of wavefunctions.

Including Spin-orbit Interaction

In a real atom the spin interacts with the magnetic field created by the electron movement around the nucleus, a phenomenon known as spin-orbit interaction. When one takes this into account, the spin and angular momentum are no longer conserved, which can be pictured by the electron precessing. Therefore, one has to replace the quantum numbers l, m and the projection of the spin m_s by quantum numbers that represent the total angular momentum (including spin), j and m_j, as well as the quantum number of parity.

The next section on the Dirac equation for a solution that includes the coupling.

Solution to Dirac Equation

In 1928 in England Paul Dirac found an equation that was fully compatible with Special Relativity. The equation was solved for hydrogen-like atoms the same year by the German Walter Gordon. Instead of a single (possibly complex) function as in the Schrödinger equation, one must find four complex functions that make up a bispinor. The first and second functions (or components of the spinor) correspond (in the usual basis) to spin "up" and spin "down" positive-energy states, whereas the third and fourth correspond to spin up and spin down negative-energy states.

The terms "spin up" and "spin down" are relative to a chosen direction, conventionally the z direction. An electron may be in a superposition of spin up and spin down, which corresponds to the spin axis pointing in some other direction. The spin state may depend on location.

An electron in the vicinity of a nucleus necessarily has non-zero amplitudes for the negative energy components. Far from the nucleus these may be small, but near the nucleus they become large.

The eigenfunctions of the Hamiltonian, which means functions with a definite energy (and which therefore do not evolve except for a phase shift), are characterized not by the quantum number n only (as for the Schrödinger equation), but by n and a quantum number j, the total angular momentum quantum number. The quantum number j determines the sum of the squares of the three angular momenta to be $j(j+1)$ (times $\hbar^2$). These angular momenta include both orbital angular momentum (having to do with the angular dependence of ψ) and spin angular momentum (having to do with the spin state). The splitting of the energies of states of the same principal quantum number n due to differences in j is called fine structure.

The orbitals for a given state can be written using two radial functions and two angle functions. The radial functions depend on both the principal quantum number n and an integer k, defined as:

$$k = \begin{cases} -j - \frac{1}{2} & \text{if } j = \ell + \frac{1}{2} \\ j + \frac{1}{2} & \text{if } j = \ell - \frac{1}{2} \end{cases}$$

where ℓ is the azimuthal quantum number that ranges from 0 to $n-1$. The angle functions depend on k and on a quantum number m which ranges from $-j/2$ to $j/2$ by steps of 1. The states are labeled using the letters S, P, D, F et cetera to stand for states with ℓ equal to 0, 1, 2, 3 et cetera, with a subscript giving j. For instance, the states for $n=4$ are given in the following table (these would be prefaced by n, for example $4S_{1/2}$):

	$m = -7/2$	$m = -5/2$	$m = -3/2$	$m = -1/2$	$m = 1/2$	$m = 3/2$	$m = 5/2$	$m = 7/2$
$k = 3, \ell = 3$		$F_{5/2}$	$F_{5/2}$	$F_{5/2}$	$F_{5/2}$	$F_{5/2}$	$F_{5/2}$	
$k = 2, \ell = 2$			$D_{3/2}$	$D_{3/2}$	$D_{3/2}$	$D_{3/2}$		
$k = 1, \ell = 1$				$P_{1/2}$	$P_{1/2}$			
$k = 0$								
$k = -1, \ell = 0$				$S_{1/2}$	$S_{1/2}$			
$k = -2, \ell = 1$			$P_{3/2}$	$P_{3/2}$	$P_{3/2}$	$P_{3/2}$		
$k = -3, \ell = 2$		$D_{5/2}$	$D_{5/2}$	$D_{5/2}$	$D_{5/2}$	$D_{5/2}$	$D_{5/2}$	
$k = -4, \ell = 3$	$F_{7/2}$	$F_{7/2}$	$F_{7/2}$	$F_{7/2}$	$F_{7/2}$	$F_{7/2}$	$F_{7/2}$	$F_{7/2}$

These can be additionally labeled with a subscript giving m. There are $2n^2$ states with principal quantum number n, $4j+2$ of them with any allowed j except the highest ($j=n-1/2$) for which there are only $2j+1$. Since the orbitals having given values of n and j have the same energy according to the Dirac equation, they form a basis for the space of functions having that energy.

The energy, as a function of n and $|k|$ (equal to $j+1/2$), is:

$$E_{nj} = \mu c^2 \left(1 + \left[\frac{Z\alpha}{n - |k| + \sqrt{k^2 - Z^2\alpha^2}} \right]^2 \right)^{-1/2}$$

$$\approx \mu c^2 \left\{ 1 - \frac{Z^2\alpha^2}{2n^2} \left[1 + \frac{Z^2\alpha^2}{n} \left(\frac{1}{|k|} - \frac{3}{4n} \right) \right] \right\}$$

(The energy of course depends on the zero-point used.) The Schrödinger solution corresponds to replacing the inner bracket in the second expression by 1. The accuracy of the energy difference between the lowest two hydrogen states calculated from the Schrödinger solution is about 9 ppm (90 µeV too low, out of around 10 eV), whereas the accuracy of the Dirac equation for the same energy difference is about 3 ppm (too high). The Schrödinger solution always puts the states at slightly higher energies than the more accurate Dirac equation. The Dirac equation gives some levels of hydrogen quite accurately (for instance the $4P_{1/2}$ state is given an energy only about 2×10^{-10} eV too high), others less so (for instance, the $2S_{1/2}$ level is about 4×10^{-6} eV too low). The modifications of the energy due to using the Dirac equation rather than the Schrödinger solution is of the order of α^2, and for this reason α is called the fine structure constant.

The solution to the Dirac equation for quantum numbers n, k, and m, is:

$$\Psi = \begin{pmatrix} g_{n,k}(r)r^{-1}\Omega_{k,m}(\theta,\phi) \\ if_{n,k}(r)r^{-1}\Omega_{-k,m}(\theta,\phi) \end{pmatrix} = \begin{pmatrix} g_{n,k}(r)r^{-1}\sqrt{(k+\frac{1}{2}-m)/(2k+1)}Y_{k,m-1/2}(\theta,\phi) \\ -g_{n,k}(r)r^{-1}\operatorname{sgn}k\sqrt{(k+\frac{1}{2}+m)/(2k+1)}Y_{k,m+1/2}(\theta,\phi) \\ if_{n,k}(r)r^{-1}\sqrt{(-k+\frac{1}{2}-m)/(-2k+1)}Y_{-k,m-1/2}(\theta,\phi) \\ -if_{n,k}(r)r^{-1}\operatorname{sgn}k\sqrt{(-k+\frac{1}{2}-m)/(-2k+1)}Y_{-k,m+1/2}(\theta,\phi) \end{pmatrix}$$

where the Ωs are columns of the two spherical harmonics functions shown to the right. $Y_{a,b}(\theta,\phi)$ signifies a spherical harmonic function:

$$Y_{a,b} = \begin{cases} (-1)^b\sqrt{\dfrac{2a+1}{4\pi}\dfrac{(a-b)!}{(a+b)!}}P_a^b(\cos\theta)e^{ib\phi} & \text{if } a > 0 \\ \\ Y_{-a-1,b} & \text{if } a < 0 \end{cases}$$

in which P_a^b is an associated Legendre polynomial. (Note that the definition of Ω may involve a spherical harmonic that doesn't exist, like $Y_{0,1}$, but the coefficient on it will be zero.)

Here is the behavior of some of these angular functions. The normalization factor is left out, and the function is multiplied by r^l to *simplify the expressions*.

$$\Omega_{-1,-1/2} \propto \begin{pmatrix} 0 \\ 1 \end{pmatrix}$$

$$\Omega_{-1,1/2} \propto \begin{pmatrix} 1 \\ 0 \end{pmatrix}$$

$$\Omega_{1,-1/2} \propto \begin{pmatrix} (x-iy)/r \\ z/r \end{pmatrix}$$

$$\Omega_{1,1/2} \propto \begin{pmatrix} z/r \\ (x+iy)/r \end{pmatrix}$$

From these we see that in the $S_{1/2}$ orbital ($k = -1$), the top two components of Ψ have zero orbital angular momentum like Schrödinger S orbitals, but the bottom two components (of negative energy) are orbitals like the Schrödinger P orbitals. In the $P_{1/2}$ solution ($k = 1$), the situation is reversed. In both cases, the spin of each component compensates for its orbital angular momentum around the z axis to give the right value for the total angular momentum around the z axis.

The two Ω spinors obey the relationship:

$$\Omega_{k,m} = \begin{pmatrix} z/r & (x-iy)/r \\ (x+iy)/r & -z/r \end{pmatrix} \Omega_{-k,m}$$

To write the functions $g_{n,k}(r)$ and $f_{n,k}(r)$ let us define a scaled radius ρ:

$$\rho \equiv 2Cr$$

with

$$C = \frac{\sqrt{\mu^2 c^4 - E^2}}{\hbar c}$$

where E is the energy (E_{nj}) given above. We also define γ as:

$$\gamma \equiv \sqrt{k^2 - Z^2\alpha^2}$$

When $k = -n$ (which corresponds to the highest j possible for a given n, such as $1S_{1/2}$, $2P_{3/2}$, $3D_{5/2}$...), then $g_{n,k}(r)$ and $f_{n,k}(r)$ are:

$$g_{n,-n}(r) = A(n+\gamma)\rho^\gamma e^{-\rho/2}$$
$$f_{n,-n}(r) = AZ\alpha\rho^\gamma e^{-\rho/2}$$

where A is a normalization constant involving the Gamma function:

$$A = \frac{1}{\sqrt{2n(n+\gamma)}}\sqrt{\frac{C}{\gamma\Gamma(2\gamma)}}$$

Notice that because of the factor $Z\alpha$, $f(r)$ (the negative energy part) is small compared to $g(r)$. Also notice that in this case, the energy is given by

$$E_{n,n-1/2} = \frac{\gamma}{n}\mu c^2 = \sqrt{1-\frac{Z^2\alpha^2}{n^2}}\mu c^2$$

and the radial decay constant C by

$$C = \frac{Z\alpha}{n}\frac{\mu c^2}{\hbar c}.$$

In the general case (when k is not $-n$), $g_{n,k}(r)$ and $f_{n,k}(r)$ are based on two generalized Laguerre polynomials:

$$g_{n,k}(r) = A\rho^\gamma e^{-\rho/2}\left(Z\alpha\rho L_{n-|k|-1}^{2\gamma+1}(\rho) + (\gamma-k)\frac{\gamma\mu c^2 - kE}{\hbar cC}L_{n-|k|}^{2\gamma-1}(\rho)\right)$$

$$f_{n,k}(r) = A\rho^\gamma e^{-\rho/2}\left((\gamma-k)\rho L_{n-|k|-1}^{2\gamma+1}(\rho) + Z\alpha\frac{\gamma\mu c^2 - kE}{\hbar cC}L_{n-|k|}^{2\gamma-1}(\rho)\right)$$

with A now defined as

$$A = \frac{1}{\sqrt{2k(k-\gamma)}} \sqrt{\frac{C}{n-|k|+\gamma} \frac{(n-|k|-1)!}{\Gamma(n-|k|+2\gamma+1)} \frac{1}{2}\left(\left(\frac{Ek}{\gamma\mu c^2}\right)^2 + \frac{Ek}{\gamma\mu c^2}\right)}$$

Again f is small compared to g (except at very small r) because when k is positive the first terms dominate, and α is big compared to $\gamma-k$, whereas when k is negative the second terms dominate and α is small compared to $\gamma-k$. Note that the dominant term is quite similar to corresponding the Schrödinger solution — the upper index on the Laguerre polynomial is slightly less ($2\gamma+1$ or $2\gamma-1$ rather than $2\ell+1$, which is the nearest integer), as is the power of ρ (γ or $\gamma-1$ instead of ℓ, the nearest integer). The exponential decay is slightly faster than in the Schrödinger solution.

The normalization factor makes the integral over all space of the square of the absolute value equal to 1.

1S Orbital

Here is the $1S_{1/2}$ orbital, spin up, without normalization:

$$\Psi \propto \begin{pmatrix} (1+\gamma)r^{\gamma-1}e^{-Cr} \\ 0 \\ iZ\alpha r^{\gamma-1}e^{-Cr}z/r \\ iZ\alpha r^{\gamma-1}e^{-Cr}(x+iy)/r \end{pmatrix}$$

Note that γ is a little less than 1, so the top function is similar to an exponentially decreasing function of r except that at very small r it theoretically goes to infinity (but this behavior appears at a value of r smaller than the radius of a proton!).

The $1S_{1/2}$ orbital, spin down, without normalization, comes out as:

$$\Psi \propto \begin{pmatrix} 0 \\ (1+\gamma)r^{\gamma-1}e^{-Cr} \\ iZ\alpha r^{\gamma-1}e^{-Cr}(x-iy)/r \\ -iZ\alpha r^{\gamma-1}e^{-Cr}z/r \end{pmatrix}$$

We can mix these in order to obtain orbitals with the spin oriented in some other direction, such as:

$$\Psi \propto \begin{pmatrix} (1+\gamma)r^{\gamma-1}e^{-Cr} \\ (1+\gamma)r^{\gamma-1}e^{-Cr} \\ iZ\alpha r^{\gamma-1}e^{-Cr}(x-iy+z)/r \\ iZ\alpha r^{\gamma-1}e^{-Cr}(x+iy-z)/r \end{pmatrix}$$

which corresponds to the spin and angular momentum axis pointing in the x direction. Adding i times the "down" spin to the "up" spin gives an orbital oriented in the y direction.

$2P_{1/2}$ and $2S_{1/2}$ Orbitals

To give another example, the $2P_{1/2}$ orbital, spin up, is proportional to:

$$\Psi \propto \begin{pmatrix} \rho^{\gamma-1}e^{-\rho/2}\left(Z\alpha\rho+(\gamma-1)\dfrac{\gamma\mu c^2-E}{\hbar cC}(-\rho+2\gamma) \right)z/r \\ \rho^{\gamma-1}e^{-\rho/2}\left(Z\alpha\rho+(\gamma-1)\dfrac{\gamma\mu c^2-E}{\hbar cC}(-\rho+2\gamma) \right)(x+iy)/r \\ i\rho^{\gamma-1}e^{-\rho/2}\left((\gamma-1)\rho+Z\alpha\dfrac{\gamma\mu c^2-E}{\hbar cC}(-\rho+2\gamma) \right) \\ 0 \end{pmatrix}$$

(Remember that $\rho = 2rC$. C is about half what it is for the 1S orbital, but γ is still the same.)

Notice that when ρ is small compared to α (or r is small compared to $\hbar c/(\mu c^2)$) the negative-energy "S" type orbital dominates (the third component of the bispinor).

For the $2S_{1/2}$ spin up orbital, we have:

$$\Psi \propto \begin{pmatrix} \rho^{\gamma-1}e^{-\rho/2}\left(Z\alpha\rho+(\gamma+1)\dfrac{\gamma\mu c^2+E}{\hbar cC}(-\rho+2\gamma) \right) \\ 0 \\ i\rho^{\gamma-1}e^{-\rho/2}\left((\gamma+1)\rho+Z\alpha\dfrac{\gamma\mu c^2+E}{\hbar cC}(-\rho+2\gamma) \right)z/r \\ i\rho^{\gamma-1}e^{-\rho/2}\left((\gamma+1)\rho+Z\alpha\dfrac{\gamma\mu c^2+E}{\hbar cC}(-\rho+2\gamma) \right)(x+iy)/r \end{pmatrix}$$

Now the positive-energy part is S-like and there is a radius near $\rho = 2$ where it goes to zero, whereas the negative-energy part is P-like.

Beyond the Dirac Equation

The Dirac equation was not the last word, and its predictions differ from experimental results as mentioned earlier. More accurate results include the Lamb shift (radiative corrections arising from quantum electrodynamics) and hyperfine structure.

References

- Peter Atkins, Julio de Paula, Ronald Friedman (2009). Quanta, Matter, and Change: A Molecular Approach to Physical Chemistry. Oxford University Press. p. 106. ISBN 978-0-19-920606-3.

- Feynman, Richard; Leighton, Robert B.; Sands, Matthew (2006). The Feynman Lectures on Physics -The Definitive Edition, Vol 1 lect 6. Pearson PLC, Addison Wesley. p. 11. ISBN 0-8053-9046-4.

- Claude Cohen-Tannoudji; Bernard Diu; Franck Laloë; et al. (1996). Quantum mechanics. Translated by from the French by Susan Reid Hemley. Wiley-Interscience. ISBN 978-0-471-56952-7.

- Catherine E. Housecroft, Alan G, Sharpe, Inorganic Chemistry, Pearson Prentice Hall; 2nd Edition, 2005, ISBN 0130-39913-2, p. 41-43.

- Clayden, Jonathan; Greeves, Nick; Warren, Stuart; Wothers, Peter (2001). Organic Chemistry (1st ed.). Oxford University Press. pp. 96–103. ISBN 978-0-19-850346-0.

- Keeler, James; Wothers, Peter (2014). Chemical Structure and Reactivity - an integrated approach (2nd ed.). Oxford University Press. pp. 123–126. ISBN 978-0-19-9604135.

- Douglas, Bodie; McDaniel, Darl; Alexander, John (1994). Concepts and Models of Inorganic Chemistry (3rd ed.). Wiley. pp. 157–159. ISBN 978-0-471-62978-8.

- Gerald Teschl (2009). Mathematical Methods in Quantum Mechanics; With Applications to Schrödinger Operators. American Mathematical Society. ISBN 978-0-8218-4660-5.

- Tipler, Paul & Ralph Llewellyn (2003). Modern Physics (4th ed.). New York: W. H. Freeman and Company. ISBN 0-7167-4345-0

Perturbation Theory and Variational Method

The Schrödinger equation in quantum chemistry is solved by approximate methods and this chapter discusses the two primary methods- perturbation theory and the variational method. The perturbation theory uses the reference of an unperturbed Hamiltonian to approximately explain the perturbation. But when it becomes difficult to pin down an unperturbed reference Hamiltonian, the variational method employs the use of a trial wavefunction. This chapter illustrates these methods in-depth.

Variational Method (Quantum Mechanics)

In quantum mechanics, the variational method is one way of finding approximations to the lowest energy eigenstate or ground state, and some excited states. This allows calculating approximate wavefunctions such as molecular orbitals. The basis for this method is the variational principle.

The method consists in choosing a "trial wavefunction" depending on one or more parameters, and finding the values of these parameters for which the expectation value of the energy is the lowest possible. The wavefunction obtained by fixing the parameters to such values is then an approximation to the ground state wavefunction, and the expectation value of the energy in that state is an upper bound to the ground state energy. The Hartree–Fock method and the Ritz method both apply the variational method.

Description

Suppose we are given a Hilbert space and a Hermitian operator over it called the Hamiltonian H. Ignoring complications about continuous spectra, we look at the discrete spectrum of H and the corresponding eigenspaces of each eigenvalue λ:

$$\sum_{\lambda_1, \lambda_2 \in \mathrm{Spec}(H)} \langle \psi_{\lambda_1} | \psi_{\lambda_2} \rangle = \delta_{\lambda_1 \lambda_2}$$

where $\delta_{i,j}$ is the Kronecker delta

$$\hat{H} |\psi_\lambda\rangle = \lambda |\psi_\lambda\rangle.$$

Physical states are normalized, meaning that their norm is equal to 1. Once again ignoring complications involved with a continuous spectrum of H, suppose it is bounded from below and that its greatest lower bound is E_0. Suppose also that we know the corresponding state $|\psi>$. The expectation value of H is then

$$\langle \psi | H | \psi \rangle \quad = \sum_{\lambda_1,\lambda_2 \in \text{Spec}(H)} \langle \psi | \psi_{\lambda_1} \rangle \langle \psi_{\lambda_1} | H | \psi_{\lambda_2} \rangle \langle \psi_{\lambda_2} | \psi \rangle$$

$$= \sum_{\lambda \in \text{Spec}(H)} \lambda \left| \langle \psi_\lambda | \psi \rangle \right|^2 \geq \sum_{\lambda \in \text{Spec}(H)} E_0 \left| \langle \psi_\lambda | \psi \rangle \right|^2 = E_0$$

Obviously, if we were to vary over all possible states with norm 1 trying to minimize the expectation value of H, the lowest value would be E_0 and the corresponding state would be an eigenstate of E_0. Varying over the entire Hilbert space is usually too complicated for physical calculations, and a subspace of the entire Hilbert space is chosen, parametrized by some (real) differentiable parameters α_i ($i = 1, 2, ..., N$). The choice of the subspace is called the ansatz. Some choices of ansatzes lead to better approximations than others, therefore the choice of ansatz is important.

Let's assume there is some overlap between the ansatz and the ground state (otherwise, it's a bad ansatz). We still wish to normalize the ansatz, so we have the constraints

$$\langle \psi(\alpha_i) | \psi(\alpha_i) \rangle = 1$$

and we wish to minimize

$$\varepsilon(\alpha_i) = \langle \psi(\alpha_i) | H | \psi(\alpha_i) \rangle.$$

This, in general, is not an easy task, since we are looking for a global minimum and finding the zeroes of the partial derivatives of ε over α_i is not sufficient. If $\psi\ (\alpha_i)$ is expressed as a linear combination of other functions (α_i being the coefficients), as in the Ritz method, there is only one minimum and the problem is straightforward. There are other, non-linear methods, however, such as the Hartree–Fock method, that are also not characterized by a multitude of minima and are therefore comfortable in calculations.

There is an additional complication in the calculations described. As ε tends toward E_0 in minimization calculations, there is no guarantee that the corresponding trial wavefunctions will tend to the actual wavefunction. This has been demonstrated by calculations using a modified harmonic oscillator as a model system, in which an exactly solvable system is approached using the variational method. A wavefunction different from the exact one is obtained by use of the method described above.

Although usually limited to calculations of the ground state energy, this method can be applied in certain cases to calculations of excited states as well. If the ground state wavefunction is known, either by the method of variation or by direct calculation, a subset of the Hilbert space can be chosen which is orthogonal to the ground state wavefunction.

$$|\psi\rangle = |\psi_{test}\rangle - \langle\psi_{gr}|\psi_{test}\rangle|\psi_{gr}\rangle$$

The resulting minimum is usually not as accurate as for the ground state, as any difference between the true ground state and ψ_{gr} results in a lower excited energy. This defect is worsened with each higher excited state.

In another formulation:

$$E_{ground} \leq \langle\phi|H|\phi\rangle.$$

This holds for any trial φ since, by definition, the ground state wavefunction has the lowest energy, and any trial wavefunction will have energy greater than or equal to it.

Proof: φ can be expanded as a linear combination of the actual eigenfunctions of the Hamiltonian (which we assume to be normalized and orthogonal):

$$\phi = \sum_n c_n \psi_n.$$

Then, to find the expectation value of the Hamiltonian:

$$\langle\phi|H|\phi\rangle$$
$$= \left\langle \sum_n c_n\psi_n \,|\, H \,|\, \sum_m c_m\psi_m \right\rangle$$
$$= \sum_n \sum_m \langle c_n\psi_n | E_m | c_m\psi_m \rangle$$
$$= \sum_n \sum_m c_n^* c_m E_m \langle\psi_n|\psi_m\rangle$$
$$= \sum_n |c_n|^2 E_n.$$

Now, the ground state energy is the lowest energy possible, i.e. $E_n \geq E_g$. Therefore, if the guessed wave function φ is normalized:

$$\langle\phi|H|\phi\rangle \geq E_g \sum_n |c_n|^2 = E_g.$$

In General

For a hamiltonian H that describes the studied system and *any* normalizable function Ψ with arguments appropriate for the unknown wave function of the system, we define the functional

$$\varepsilon[\Psi] = \frac{\langle \Psi | \hat{H} | \Psi \rangle}{\langle \Psi | \Psi \rangle}.$$

The variational principle states that

- $\varepsilon \geq E_0$, where E_0 is the lowest energy eigenstate (ground state) of the hamiltonian

- $\varepsilon = E_0$ if and only if Ψ is exactly equal to the wave function of the ground state of the studied system.

The variational principle formulated above is the basis of the variational method used in quantum mechanics and quantum chemistry to find approximations to the ground state.

Another facet in variational principles in quantum mechanics is that since Ψ and $\Psi^{\dagger}$ can be varied separately (a fact arising due to the complex nature of the wave function), the quantities can be varied in principle just one at a time.

Helium Atom Ground State

The helium atom consists of two electrons with mass m and electric charge $-e$, around an essentially fixed nucleus of mass $M \gg m$ and charge $+2e$. The Hamiltonian for it, neglecting the fine structure, is:

$$H = -\frac{\hbar^2}{2m}(\nabla_1^2 + \nabla_2^2) - \frac{e^2}{4\pi \epsilon_0}\left(\frac{2}{r_1} + \frac{2}{r_2} - \frac{1}{|\mathbf{r}_1 - \mathbf{r}_2|}\right)$$

where $\hbar$ is the reduced Planck constant, ε_0 is the vacuum permittivity, r_i (for $i = 1, 2$) is the distance of the i-th electron from the nucleus, and $|\mathbf{r}_1 - \mathbf{r}_2|$ is the distance between the two electrons.

If the term $V_{ee} = e^2/(4\pi\varepsilon_0|\mathbf{r}_1 - \mathbf{r}_2|)$, representing the repulsion between the two electrons, were excluded, the Hamiltonian would become the sum of two hydrogen-like atom Hamiltonians with nuclear charge $+2e$. The ground state energy would then be $8E_1 = -109$ eV, where E_1 is the Rydberg constant, and its ground state wavefunction would be the product of two wavefunctions for the ground state of hydrogen-like atoms:

$$\psi(\mathbf{r}_1, \mathbf{r}_2) = \frac{Z^3}{\pi a_0^3} e^{-Z(r_1 + r_2)/a_0}.$$

where a_0 is the Bohr radius and $Z = 2$, helium's nuclear charge. The expectation value of the total Hamiltonian H (including the term V_{ee}) in the state described by ψ_0 will be an upper bound for its ground state energy. $<V_{ee}>$ is $-5E_1/2 = 34$ eV, so $<H>$ is $8E_1 - 5E_1/2 = -75$ eV.

A tighter upper bound can be found by using a better trial wavefunction with 'tunable' parameters. Each electron can be thought to see the nuclear charge partially "shielded" by the other electron, so we can use a trial wavefunction equal with an "effective" nuclear charge $Z < 2$: The expectation value of H in this state is:

$$\langle H \rangle = \left[-2Z^2 + \frac{27}{4} Z \right] E_1$$

This is minimal for $Z = 27/16$; Shielding reduces the effective charge to ~1.69. Substituting this value of Z into the expression for H yields $729E_1/128 = -77.5$ eV, within 2% of the experimental value, -78.975 eV.

Even closer estimations of this energy have been found using more complicated trial wave functions with more parameters. This is done in physical chemistry via variational Monte Carlo.

Perturbation Theory

Perturbation theory comprises mathematical methods for finding an approximate solution to a problem, by starting from the exact solution of a related, simpler problem. A critical feature of the technique is a middle step that breaks the problem into "solvable" and "perturbation" parts. Perturbation theory is applicable if the problem at hand cannot be solved exactly, but can be formulated by adding a "small" term to the mathematical description of the exactly solvable problem.

Perturbation theory leads to an expression for the desired solution in terms of a formal power series in some "small" parameter – known as a perturbation series – that quantifies the deviation from the exactly solvable problem. The leading term in this power series is the solution of the exactly solvable problem, while further terms describe the deviation in the solution, due to the deviation from the initial problem. Formally, we have for the approximation to the full solution A, a series in the small parameter (here called ε), like the following:

$$A = A_0 + \varepsilon^1 A_1 + \varepsilon^2 A_2 + \cdots$$

In this example, A_0 would be the known solution to the exactly solvable initial problem and $A_1, A_2, \ldots$ represent the higher-order terms which may be found iteratively by some systematic procedure. For small ε these higher-order terms in the series become successively smaller.

An approximate "perturbation solution" is obtained by truncating the series, usually by keeping only the first two terms, the initial solution and the "first-order" perturbation correction

$$A \approx A_0 + \varepsilon A_1 .$$

General Description

Perturbation theory is closely related to methods used in numerical analysis. The earliest use of what would now be called *perturbation theory* was to deal with the otherwise unsolvable mathematical problems of celestial mechanics: for example the orbit of the Moon, which moves noticeably differently from a simple Keplerian ellipse because of the competing gravitation of the Earth and the Sun.

Perturbation methods start with a simplified form of the original problem, which is *simple enough* to be solved exactly. In celestial mechanics, this is usually a Keplerian ellipse. Under non-relativistic gravity, an ellipse is exactly correct when there are only two gravitating bodies (say, the Earth and the Moon) but not quite correct when there are three or more objects (say, the Earth, Moon, Sun, and the rest of the solar system) and not quite correct when the gravitational interaction is stated using formulas from General relativity.

The solved, but simplified problem is then *"perturbed"* to make the conditions that the perturbed solution actually satisfies closer to the real problem, such as including the gravitational attraction of a third body (the Sun). The "conditions" are a formula (or several) that represent reality, often something arising from a physical law like Newton's second law, the force-acceleration equation,

$$\mathbf{F} = m\mathbf{a} .$$

In the case of the example, the force $\mathbf{F}$ is calculated based on the number of gravitationally relevant bodies; the acceleration $\mathbf{a}$ is obtained, using calculus, from the path of the Moon in its orbit. Both of these come in two forms: approximate values for force and acceleration, which result from simplifications, and hypothetical exact values for force and acceleration, which would require the complete answer to calculate.

The slight changes that result from accommodating the perturbation, which themselves may have been simplified yet again, are used as corrections to the approximate solution. Because of simplifications introduced along every step of the way, the corrections are never perfect, and the conditions met by the corrected solution do not perfectly match the equation demanded by reality. However, even only one cycle of corrections often provides an excellent approximate answer to what the real solution should be.

There is no requirement to stop at only one cycle of corrections. A partially corrected solution can be re-used as the new starting point for yet another cycle of perturbations and corrections. In principle, cycles of finding increasingly better corrections could go on indefinitely. In practice, one typically stops at one or two cycles of corrections. The usual difficulty with the method is that the corrections progressively make the new solutions very much more complicated, so each cycle is much more difficult to manage than the previous cycle of corrections. Isaac Newton is reported to have said, regarding the problem of the Moon's orbit, that *"It causeth my head to ache."*

This general procedure is a widely used mathematical tool in advanced sciences and engineering: start with a simplified problem and gradually add corrections that make the formula that the corrected problem matches closer and closer to the formula that represents reality. It is the natural extension to mathematical functions of the "guess, check, and fix" method used by older civilisations to compute certain numbers, such as square roots.

Examples

Examples for the "mathematical description" are: an algebraic equation, a differential equation (e.g., the equations of motion in celestial mechanics or a wave equation), a free energy (in statistical mechanics), a Hamiltonian operator (in quantum mechanics).

Examples for the kind of solution to be found perturbatively: the solution of the equation (e.g., the trajectory of a particle), the statistical average of some physical quantity (e.g., average magnetization), the ground state energy of a quantum mechanical problem.

Examples for the exactly solvable problems to start with: linear equations, including linear equations of motion (harmonic oscillator, linear wave equation), statistical or quantum-mechanical systems of non-interacting particles (or in general, Hamiltonians or free energies containing only terms quadratic in all degrees of freedom).

Examples of "perturbations" to deal with: Nonlinear contributions to the equations of motion, interactions between particles, terms of higher powers in the Hamiltonian/ Free Energy.

For physical problems involving interactions between particles, the terms of the perturbation series may be displayed (and manipulated) using Feynman diagrams.

History

Perturbation theory was first devised to solve otherwise intractable problems in the calculation of the motions of planets in the solar system. The gradually increasing accuracy of astronomical observations led to incremental demands in the accuracy of solutions to Newton's gravitational equations, which led several notable 18th and 19th century mathematicians to extend and generalize the methods of perturbation theory. These well-developed perturbation methods were adopted and adapted to solve new problems arising during the development of Quantum Mechanics in 20th century atomic and subatomic physics.

Beginnings In the Study of Planetary Motion

Since the planets are very remote from each other, and since their mass is small as compared to the mass of the Sun, the gravitational forces between the planets can be neglected, and the planetary motion is considered, to a first approximation, as taking place along Kepler's orbits, which are defined by the equations of the two-body problem, the two bodies being the planet and the Sun.

Since astronomic data came to be known with much greater accuracy, it became necessary to consider how the motion of a planet around the Sun is affected by other planets. This was the origin of the three-body problem; thus, in studying the system Moon–Earth–Sun the mass ratio between the Moon and the Earth was chosen as the small parameter. Lagrange and Laplace were the first to advance the view that the constants which describe the motion of a planet around the Sun are "perturbed", as it were, by the motion of other planets and vary as a function of time; hence the name "perturbation theory".

Perturbation theory was investigated by the classical scholars — Laplace, Poisson, Gauss — as a result of which the computations could be performed with a very high accuracy. The discovery of the planet Neptune in 1848 by Urbain Le Verrier, based on the deviations in motion of the planet Uranus (he sent the coordinates to Johann Gottfried Galle who successfully observed Neptune through his telescope), represented a triumph of perturbation theory.

Rise of Understanding of Chaotic Systems

The development of basic perturbation theory for differential equations was fairly complete by the middle of the 19th century. It was at that time that Charles-Eugène Delaunay was studying the perturbative expansion for the Earth-Moon-Sun system, and discovered the so-called "problem of small denominators". Here, the denominator appearing in the n-th term of the perturbative expansion could become arbitrarily small, causing the n-th correction to be as large or larger than the first-order correction.

At the turn of the 20th century, this problem led Henri Poincaré to make one of the first deductions of the existence of chaos,and what is poetically called the "butterfly effect":that even a very small perturbation can ultimately have a very large effect on non-dissipative or "friction-free" dynamic systems.

A partial resolution of the small-divisor problem was given by the statement of the KAM theorem in 1954. Developed by Andrey Kolmogorov, Vladimir Arnold and Jürgen Moser, this theorem stated the conditions under which a system of partial differential equations will have only mildly chaotic behaviour under small perturbations.

Application to New Problems in 20th Century Physics

Perturbation theory saw a particularly dramatic expansion and evolution with the arrival of quantum mechanics. Although perturbation theory was used in the semi-classical theory of the Bohr atom, the calculations were monstrously complicated, and subject to somewhat ambiguous interpretation. The discovery of Heisenberg's matrix mechanics allowed a vast simplification of the application of perturbation theory. Notable examples are the Stark effect and the Zeeman effect, which have a simple enough theory to be included in standard undergraduate textbooks in quantum mechanics. Other early applications include the fine structure and the hyperfine structure in the hydrogen atom.

In modern times, perturbation theory underlies much of quantum chemistry and quantum field theory. In chemistry, perturbation theory was used to obtain the first solutions for the helium atom.

In the middle of the 20th century, Richard Feynman realized that the perturbative expansion could be given a dramatic and beautiful graphical representation in terms of what are now called Feynman diagrams. Although originally applied only in quantum field theory, such diagrams now find increasing use in any area where perturbative expansions are studied.

Search for Better Methods for Quantum Mechanics

In the late 20th century, broad dissatisfaction with perturbation theory in the quantum physics community, including not only the difficulty of going beyond second order in the expansion, but also questions about whether the perturbative expansion is even convergent, has led to a strong interest in the area of non-perturbative analysis, that is, the study of exactly solvable models.

Much of the theoretical work in non-perturbative analysis goes under the name of quantum groups and non-commutative geometry. The prototypical model is the Korteweg–de Vries equation, a highly non-linear equation for which the interesting solutions, the solitons, cannot be reached by perturbation theory, even if the perturbations were carried out to infinite order.

Perturbation Orders

The standard exposition of perturbation theory is given in terms of the *order* to which the perturbation is carried out: first-order perturbation theory or second-order perturbation theory, and whether the perturbed states are degenerate, which requires singular perturbation. In the singular case extra care must be taken, and the theory is slightly more elaborate.

First-order, Non-singular Perturbation Theory

This section develops, in simple terms, the general theory for the perturbative solution to a differential equation to the first order. To keep the exposition simple, a crucial assumption is made: that the solutions to the unperturbed system are not *degenerate*, so that the perturbation series can be inverted. There are ways of dealing with the degenerate (or *singular*) case; these require extra care.

Suppose one wants to solve a differential equation of the form

$$Dg(x) = \lambda g(x),$$

where D is some specific differential operator, and λ is an eigenvalue. Many problems involving ordinary or partial differential equations can be cast in this form.

It is presumed that the differential operator can be written in the form

$$D = D^{(0)} + \varepsilon D^{(1)}$$

where ε is presumed to be small, and that, furthermore, the complete set of solutions for $D^{(0)}$ are known.

That is, one has a set of solutions $f_n^{(0)}(x)$, labelled by some arbitrary index n, such that

$$D^{(0)} f_n^{(0)}(x) = \lambda_n^{(0)} f_n^{(0)}(x).$$

Furthermore, one assumes that the set of solutions $\{f_n^{(0)}(x)\}$ form an orthonormal set,

$$\int f_m^{(0)}(x) f_n^{(0)}(x) dx = \delta_{mn}$$

with δ_{mn} the Kronecker delta function.

To zeroth order, one expects that the solutions $g(x)$ are then somehow "close" to one of the unperturbed solutions $f_n^{(0)}(x)$. That is,

$$g(x) = f_n^{(0)}(x) + \mathcal{O}(\varepsilon)$$

and

$$\lambda = \lambda_n^{(0)} + \mathcal{O}(\varepsilon).$$

where $\mathcal{O}$ denotes the relative size, in big-O notation, of the perturbation.

To solve this problem, one assumes that the solution $g(x)$ can be written as a linear combination of the $f_n^{(0)}(x)$,

$$g(x) = \sum_m c_m f_m^{(0)}(x)$$

with all of the constants $c_m = \mathcal{O}(\varepsilon)$ except for n, where $c_n = \mathcal{O}(1)$.

Substituting this last expansion into the differential equation, taking the inner product of the result with $f_n^{(0)}(x)$, and making use of orthogonality, one obtains

$$c_n \lambda_n^{(0)} + \varepsilon \sum_m c_m \int f_n^{(0)}(x) D^{(1)} f_m^{(0)}(x) dx = \lambda c_n.$$

This can be trivially rewritten as a simple linear algebra problem of finding the eigenvalue of a matrix, where

$$\sum_m A_{nm} c_m = \lambda c_n$$

where the matrix elements A_{nm} are given by

$$A_{nm} = \delta_{nm} \lambda_n^{(0)} + \varepsilon \int f_n^{(0)}(x) D^{(1)} f_m^{(0)}(x) dx \ .$$

Rather than solving this full matrix equation, one notes that, of all the c_m in the linear equation, only one, namely c_n, is not small. Thus, to the *first order* in ε, the linear equation may be solved trivially as

$$\lambda = \lambda_n^{(0)} + \varepsilon \int f_n^{(0)}(x) D^{(1)} f_n^{(0)}(x) dx$$

since all of the other terms in the linear equation are of order $\mathcal{O}(\varepsilon^2)$. The above gives the solution of the eigenvalue to first order in perturbation theory.

The function $g(x)$ to first order is obtained through similar reasoning. Substituting

$$g(x) = f_n^{(0)}(x) + \varepsilon f_n^{(1)}(x)$$

so that

$$\left(D^{(0)} + \varepsilon D^{(1)} \right) \left(f_n^{(0)}(x) + \varepsilon f_n^{(1)}(x) \right) = \left(\lambda_n^{(0)} + \varepsilon \lambda_n^{(1)} \right) \left(f_n^{(0)}(x) + \varepsilon f_n^{(1)}(x) \right)$$

gives an equation for $f_n^{(1)}(x)$.

It may be solved integrating with the partition of unity

$$\delta(x - y) = \sum_n f_n^{(0)}(x) f_n^{(0)}(y)$$

to give

$$f_n^{(1)}(x) = \sum_{m(\neq n)} \frac{f_m^{(0)}(x)}{\lambda_n^{(0)} - \lambda_m^{(0)}} \int f_m^{(0)}(y) D^{(1)} f_n^{(0)}(y) dy$$

which finally gives the exact solution to the perturbed differential equation to first order in the perturbation ε.

Several observations may be made about the form of this solution. First, the sum over functions with differences of eigenvalues in the denominator evokes the resolvent in Fredholm theory. This is no accident; the resolvent acts essentially as a kind of Green's function or propagator, passing the perturbation along. Higher-order perturbations resemble this form, with an additional sum over a resolvent appearing at each order.

The form of this solution also illustrates the idea behind the small-divisor problem. If, for whatever reason, two eigenvalues are close, so that the difference $\lambda_n^{(0)} - \lambda_m^{(0)}$ becomes small, the corresponding term in the above sum will become disproportionately large. In particular, if this happens in higher-order terms, the higher-order perturbation may become as large or larger in magnitude than the first-order perturbation. Such a situation calls into question the validity of utilizing a perturbative analysis to begin with, which can be understood to be a fairly catastrophic situation; it is frequently encountered in chaotic dynamical systems, and requires the development of techniques other than perturbation theory to solve the problem.

Curiously, the situation is not at all bad if two or more eigenvalues are *exactly equal*. This case is referred to as singular or degenerate perturbation theory, addressed below. The degeneracy of eigenvalues indicates that the unperturbed system has some sort of symmetry, and that the generators of that symmetry commute with the unperturbed differential operator. Typically, the perturbing term does not possess the symmetry, and so the full solutions do not, either; one says that the perturbation *lifts* or *breaks* the degeneracy. In this case, the perturbation can still be performed, as in following sections; however, care must be taken to work in a basis for the unperturbed states, so that these map one-to-one to the perturbed states, rather than being a mixture.

Perturbation Theory of Degenerate States

One may note that a problem occurs in the above first order perturbation theory when two or more eigenfunctions of the unperturbed system correspond to the same eigenvalue, i.e. when the eigenvalue equation becomes

$$D^{(0)} f_{n,i}^{(0)}(x) = \lambda_n^{(0)} f_{n,i}^{(0)}(x) ,$$

and the index i labels *several states with the same eigenvalue* $\lambda_n^{(0)}$. The expression for the eigenfunctions which has energy differences in the denominators becomes infinite. In that case, degenerate perturbation theory must be applied.

The degeneracy must first be removed for higher order perturbation theory. First, consider the eigenfunction which is a linear combination of eigenfunctions with the same eigenvalue only,

$$g(x) = \sum_k c_{n,k} f_{n,k}^{(0)}(x) ,$$

which, again from the orthogonality of $f_{n,k}^{(0)}$, leads to the following equation,

$$c_{n,i} \lambda_n^{(0)} + \varepsilon \sum_k c_{n,k} \int f_{n,i}^{(0)}(x) D^{(1)} f_{n,k}^{(0)}(x) dx = \lambda c_{n,i}$$

for each n.

As for the majority of low quantum numbers n, i changes over a *small range of integers*, so often the later equation can be solved analytically as an at most 4×4 matrix equation. Once the degeneracy is removed, the first and any order of the above perturbation theory may be further applied relying on the new eigenfunctions.

An Example of Second-order Singular Perturbation Theory

Consider the following equation for the unknown variable x:

$$x = 1 + \varepsilon x^5.$$

For the initial problem with $\varepsilon = 0$, the solution is $x_0 = 1$. For small ε the lowest-order approximation may be found by inserting the ansatz

$$x = x_0 + \varepsilon x_1 + \cdots$$

into the equation and demanding the equation to be fulfilled up to terms that involve powers of ε higher than the first. This yields $x_1 = 1$. In the same way, the higher orders may be found. However, even in this simple example it may be observed that for (arbitrarily) small positive ε there are four other solutions to the equation (with very large magnitude). The reason we don't find these solutions in the above perturbation method is because these solutions diverge when $\varepsilon \to 0$ while the ansatz assumes regular behavior in this limit.

The four additional solutions can be found using the methods of singular perturbation theory. In this case this works as follows. Since the four solutions diverge at $\varepsilon = 0$, it makes sense to rescale x. We put

$$x = y \varepsilon^{-\nu}$$

such that in terms of y the solutions stay finite. This means that we need to choose the exponent ν to match the rate at which the solutions diverge. In terms of y the equation reads:

$$\varepsilon^{-\nu} y = 1 + \varepsilon^{1-5\nu} y^5$$

The 'right' value for ν is obtained when the exponent of ε in the prefactor of the term proportional to y is equal to the exponent of ε in the prefactor of the term proportional to y^5, i.e. when $\nu = 1/4$. This is called 'significant degeneration'. If we choose ν larger, then the four solutions will collapse to zero in terms of y and they will become degenerate with the solution we found above. If we choose ν smaller, then the four solutions will still diverge to infinity.

Putting $\nu = 1/4$ in the above equation yields:

$$y = \varepsilon^{\frac{1}{4}} + y^5$$

This equation can be solved using ordinary perturbation theory in the same way as regular expansion for x was obtained. Since the expansion parameter is now $\varepsilon^{1/4}$ we put:

$$y = y_0 + \varepsilon^{\frac{1}{4}} y_1 + \varepsilon^{\frac{1}{2}} y_2 + \cdots$$

There are five solutions for y_0: $\{0, \pm 1, \pm i\}$. We must disregard the solution $y = 0$ since it corresponds to the original regular solution which appears to be at zero for $\varepsilon = 0$, because in the limit $\varepsilon \to 0$ we are rescaling by an infinite amount. The next term is $y_1 = -1/4$. In terms of x the four solutions are thus given as:

$$x = \varepsilon^{-\frac{1}{4}} \left[y_0 - \frac{1}{4} \varepsilon^{\frac{1}{4}} + \cdots \right]$$

Example of Degenerate Perturbation Theory – Stark Effect in Resonant Rotating Wave

Let us consider a hydrogen atom rotating with a constant angular frequency ω in an electric field. The Hamiltonian is given by:

$$H = H_0 + \varepsilon x$$

where the unperturbed Hamiltonian is

$$H_0 = \frac{\mathbf{p}^2}{2} - \frac{1}{r} - \omega L_z,$$

and L_z is the operator for the z-component of angular momentum: $L_z = i\partial/\partial\varphi$. The perturbation εx can be seen as the strength of the applied electric field multiplied by one of the space coordinates (This calculation is in atomic units, so that every quantity is dimensionless).

The eigenvalues of H_0 are

$$E_{n,m} = -\frac{1}{2}n^2 - m\omega$$

For the lowest energy eigenstates of Hydrogen $|n,l,m\rangle, |1,0,0\rangle$ and $|2,1,1\rangle$ in the resonance $E_{2,1} - E_{1,0} = 0$ their energies are therefore equal $E_{1,0} = E_{2,1} = -1/2$, while the eigenstates are different.

The eigenvalue equation for the Hamiltonian takes the form

$$\begin{bmatrix} E_{1,0} & \varepsilon d \\ \varepsilon d & E_{1,0} \end{bmatrix} \begin{bmatrix} a \\ b \end{bmatrix} = E \begin{bmatrix} a \\ b \end{bmatrix}$$

where

$$d = \frac{128}{243} a_0$$

which leads to the quadratic equation which can be readily solved

$$\left(E_{1,0} - E \right)^2 - d^2 \varepsilon^2 = 0$$

with the solution

$$|\chi 1\rangle = \frac{1}{\sqrt{2}}(|1,0,0\rangle + |2,1,1\rangle)$$

$$E(1) = E_{1,0} + d\varepsilon$$

$$|\chi 2\rangle = \frac{1}{\sqrt{2}}(|1,0,0\rangle - |2,1,1\rangle)$$

$$E(2) = E_{1,0} - d\varepsilon$$

These states are the Stark states in the rotating frame, they are Trojan (higher eigenvalue) and anti-Trojan wavepackets.

Some Modern Applications and Limitations

Both regular and singular perturbation theory are frequently used in physics and engineering. Regular perturbation theory may only be used to find those solutions of a problem that evolve smoothly out of the initial solution when changing the parameter (that are "adiabatically connected" to the initial solution).

A well-known example from physics where regular perturbation theory fails is in fluid dynamics when one treats the viscosity as a small parameter. Close to a boundary, the fluid velocity goes to zero, even for very small viscosity (the no-slip condition). For zero viscosity, it is not possible to impose this boundary condition and a regular perturbative expansion amounts to an expansion about an unrealistic physical solution. Singular perturbation theory can, however, be applied here and this amounts to 'zooming in' at the boundaries (using the method of matched asymptotic expansions).

Perturbation theory can fail when the system can transition to a different "phase" of matter, with a qualitatively different behaviour, that cannot be modelled by the physical formulas put into the perturbation theory (e.g., a solid crystal melting into a liquid). In some cases, this failure manifests itself by divergent behavior of the perturbation series. Such divergent series can sometimes be resummed using techniques such as Borel resummation.

Perturbation techniques can be also used to find approximate solutions to non-linear differential equations. Examples of techniques used to find approximate solutions to these types of problems are the Lindstedt–Poincaré technique and the method of multiple time scales.

There is absolutely no guarantee that perturbative methods result in a convergent solution. In fact, asymptotic series are the norm.

Perturbation Theory in Chemistry

Many of the ab initio quantum chemistry methods use perturbation theory directly or are closely related methods. Implicit perturbation theory works with the complete Hamiltonian from the very beginning and never specifies a perturbation operator as such. Møller–Plesset perturbation theory uses the difference between the Hartree–Fock Hamiltonian and the exact non-relativistic Hamiltonian as the perturbation. The zero-order energy is the sum of orbital energies. The first-order energy is the Hartree–Fock energy and electron correlation is included at second-order or higher. Calculations to second, third or fourth order are very common and the code is included in most ab initio quantum chemistry programs. A related but more accurate method is the coupled cluster method.

Singular Perturbation

In mathematics, a singular perturbation problem is a problem containing a small parameter that cannot be approximated by setting the parameter value to zero. More precisely, the solution cannot be uniformly approximated by an asymptotic expansion

$$\varphi(x) \approx \sum_{n=0}^{N} \delta_n(\varepsilon)\psi_n(x)$$

as $\varepsilon \to 0$. Here ε is the small parameter of the problem and $\delta_n(\varepsilon)$ are a sequence of functions of ε of increasing order, such as $\delta_n(\varepsilon) = \varepsilon^n$. This is in contrast to regular perturbation problems, for which a uniform approximation of this form can be obtained. Singularly perturbed problems are generally characterized by dynamics operating on multiple scales. Several classes of singular perturbations are outlined below.

Methods of Analysis

A perturbed problem whose solution can be approximated on the whole problem domain, whether space or time, by a single asymptotic expansion has a regular perturbation. Most often in applications, an acceptable approximation to a regularly perturbed problem is found by simply replacing the small parameter by zero everywhere in the problem statement. This corresponds to taking only the first term of the expansion, yielding an approximation that converges, perhaps slowly, to the true solution as decreases. The solution to a singularly perturbed problem cannot be approximated in this way: As seen in the examples below, a singular perturbation generally occurs when a problem's small parameter multiplies its highest operator. Thus naively taking the parameter to be zero changes the very nature of the problem. In the case of differential equations, boundary conditions cannot be satisfied; in algebraic equations, the possible number of solutions is decreased.

Singular perturbation theory is a rich and ongoing area of exploration for mathematicians, physicists, and other researchers. The methods used to tackle problems in this field are many. The more basic of these include the method of matched asymptotic expansions and WKB approximation for spatial problems, and in time, the Poincaré-Lindstedt method, the method of multiple scales and periodic averaging.

For books on singular perturbation in ODE and PDE's, see for example Holmes, *Introduction to Perturbation Methods*, Hinch, *Perturbation methods* or Bender and Orszag, *Advanced Mathematical Methods for Scientists and Engineers*.

Examples of Singular Perturbative Problems

Each of the examples described below shows how a naive perturbation analysis, which assumes that the problem is regular instead of singular, will fail. Some show how the problem may be solved by more sophisticated singular methods.

Vanishing Coefficients in Ordinary Differential Equations

Differential equations that contain a small parameter that premultiplies the highest order term typically exhibit boundary layers, so that the solution evolves in two different scales. For example, consider the boundary value problem

$$\varepsilon u''(x) + u'(x) = -e^{-x}, \ 0 < x < 1$$
$$u(0) = 0, \ u(1) = 1.$$

Its solution when $\varepsilon = 0.1$ is the solid curve shown below. Note that the solution changes rapidly near the origin. If we naively set $\varepsilon = 0$, we would get the solution labelled "outer" below which does not model the boundary layer, for which x is close to zero. For more details that show how to obtain the uniformly valid approximation, see method of matched asymptotic expansions.

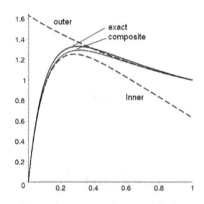

Examples in Time

An electrically driven robot manipulator can have slower mechanical dynamics and faster electrical dynamics, thus exhibiting two time scales. In such cases, we can divide the system into two subsystems, one corresponding to faster dynamics and other corresponding to slower dynamics, and then design controllers for each one of them separately. Through a singular perturbation technique, we can make these two subsystems independent of each other, thereby simplifying the control problem.

Consider a class of system described by following set of equations:

$$\dot{x}_1 = f_1(x_1, x_2) + \varepsilon g_1(x_1, x_2, \varepsilon),$$
$$\varepsilon \dot{x}_2 = f_2(x_1, x_2) + \varepsilon g_2(x_1, x_2, \varepsilon),$$
$$x_1(0) = a_1, x_2(0) = a_2,$$

with $0 < \varepsilon \ll 1$. The second equation indicates that the dynamics of x_2 is much faster than that of x_1. A theorem due to Tikhonov states that, with the correct conditions on the system, it will initially and very quickly approximate the solution to the equations

$$\dot{x}_1 = f_1(x_1, x_2),$$

$$f_2(x_1, x_2) = 0,$$

$$x_1(0) = a_1,$$

on some interval of time and that, as ε decreases toward zero, the system will approach the solution more closely in that same interval.

Examples in Space

In fluid mechanics, the properties of a slightly viscous fluid are dramatically different outside and inside a narrow boundary layer. Thus the fluid exhibits multiple spatial scales.

Reaction-diffusion systems in which one reagent diffuses much more slowly than another can form spatial patterns marked by areas where a reagent exists, and areas where it does not, with sharp transitions between them. In ecology, predator-prey models such as

$$u_t = \varepsilon u_{xx} + uf(u) - vg(u),$$

$$v_t = v_{xx} + vh(u),$$

where u is the prey and v is the predator, have been shown to exhibit such patterns.

Algebraic Equations

Consider the problem of finding all roots of the polynomial $p(x) = \varepsilon x^3 - x^2 + 1$. In the limit $\varepsilon \to 0$, this cubic degenerates into the quadratic $1 - x^2$ with roots at $x = \pm 1$.

Singular perturbation analysis suggests that the cubic has another root $x \approx 1/\varepsilon$. Indeed, with $\varepsilon = 0.1$, the roots are -0.955, 1.057, and 9.898. With $\varepsilon = 0.01$, the roots are -0.995, 1.005, and 99.990. With $\varepsilon = 0.001$, the roots are -0.9995, 1.0005, and 999.999.

In a sense, the problem has two different scales: two of the roots converge to finite numbers as ε decreases, while the third becomes arbitrarily large.

Ab Initio Quantum Chemistry Methods

Ab initio quantum chemistry methods are computational chemistry methods based on quantum chemistry. The term *ab initio* was first used in quantum chemistry by Robert Parr and coworkers, including David Craig in a semiempirical study on the excited states of benzene. The background is described by Parr. In its modern meaning ('from first principles of quantum mechanics') the term was used by Chen (when quoting an unpublished 1955 MIT report by Allen and Nesbet), by Roothaan and, in the title of an article, by Allen and Karo, who also clearly define it.

Almost always the basis set (which is usually built from the LCAO ansatz) used to solve the Schrödinger equation is not complete, and does not span the Hilbert space associat-ed with ionization and scattering processes. In the Hartree–Fock method and the configuration interaction method, this approximation allows one to treat the Schrödinger equation as a "simple" eigenvalue equation of the electronic molecular Hamiltonian, with a discrete set of solutions.

Accuracy and Scaling

Ab initio electronic structure methods have the advantage that they can be made to converge to the exact solution, when all approximations are sufficiently small in magnitude and when the finite set of basis functions tends toward the limit of a complete set. In this case, configuration interaction, where all possible configurations are included (called "Full CI"), tends to the exact non-relativistic solution of the electronic Schrödinger equation (in the Born–Oppenheimer approximation). The convergence, however, is usually not monotonic, and sometimes the smallest calculation gives the best result for some properties.

One needs to consider the computational cost of *ab initio* methods when determining whether they are appropriate for the problem at hand. When compared to much less accurate approaches, such as molecular mechanics, *ab initio* methods often take larger amounts of computer time, memory, and disk space, though, with modern advances in computer science and technology such considerations are becoming less of an issue. The HF method scales nominally as N^4 (N being a relative measure of the system size, not the number of basis functions) – e.g., if you double the number of electrons and the number of basis functions (double the system size), the calculation will take 16 (2^4) times as long per iteration. However, in practice it can scale closer to N^3 as the program can identify zero and extremely small integrals and neglect them. Correlated calculations scale less favorably, though their accuracy is usually greater, which is the trade off one needs to consider: Second–order many–body perturbation theory (MBPT(2)), or when the HF reference is used, Møller–Plesset perturbation theory (MP2) scales as N^4 or N^5, depending on how it is implemented, MP3 scales as N^6 and coupled cluster with singles and doubles (CCSD) scales iteratively as N^6, MP4 scales as N^7 and CCSD(T) and CR-CC(2,3) scale iteratively as N^6, with one noniterative step which scales as N^7. Density functional theory (DFT) methods using functionals which include Hartree–Fock exchange scale in a similar manner to Hartree–Fock but with a larger proportionality term and are thus more expensive than an equivalent Hartree–Fock calculation. DFT methods that do not include Hartree–Fock exchange can scale better than Hartree–Fock.

Linear Scaling Approaches

The problem of computational expense can be alleviated through simplification schemes. In the *density fitting* scheme, the four-index integrals used to describe the interaction between electron pairs are reduced to simpler two- or three-index integrals, by treating the charge densities they contain in a simplified way. This reduces the scaling with respect to basis set size. Methods employing this scheme are denoted by the prefix "df-", for example the density fitting MP2 is df-MP2 (many authors use lower-case to prevent confusion with DFT). In the *local approximation*, the molecular orbitals are first localized by a unitary rotation in the orbital space (which leaves the reference wave function invariant, i.e., is not an approximation) and subsequently interactions of distant pairs of localized orbitals are neglected in the correlation calculation. This sharply reduces the scaling with molecular size, a major problem in the treatment of biologically-sized molecules. Methods employing this scheme are denoted by the prefix "L", e.g. LMP2. Both schemes can be employed together, as in the df-LMP2 and df-LCCSD(T0) methods. In fact, df-LMP2 calculations are faster than df-Hartree–Fock calculations and thus are feasible in nearly all situations in which also DFT is.

Classes of Methods

The most popular classes of *ab initio* electronic structure methods:

Hartree–Fock Methods

- Hartree–Fock (HF)

- Restricted open-shell Hartree–Fock (ROHF)

- Unrestricted Hartree–Fock (UHF)

Post-Hartree–Fock Methods

- Møller–Plesset perturbation theory (MPn)

- Configuration interaction (CI)

- Coupled cluster (CC)

- Quadratic configuration interaction (QCI)

- Quantum chemistry composite methods

Multi-reference Methods

- Multi-configurational self-consistent field (MCSCF including CASSCF and RASSCF)

- Multi-reference configuration interaction (MRCI)

- n-electron valence state perturbation theory (NEVPT)

- Complete active space perturbation theory (CASPTn)

- State universal multi-reference coupled-cluster theory (SUMR-CC)

Methods in Detail

Hartree–Fock and Post-Hartree–Fock Methods

The simplest type of *ab initio* electronic structure calculation is the Hartree–Fock (HF) scheme, in which the instantaneous Coulombic electron-electron repulsion is not specifically taken into account. Only its average effect (mean field) is included in the calculation. This is a variational procedure; therefore, the obtained approximate energies, expressed in terms of the system's wave function, are always equal to or greater than the exact energy, and tend to a limiting value called the Hartree–Fock limit as the size of the basis is increased. Many types of calculations begin with a Hartree–Fock calculation and subsequently correct for electron-electron repulsion, referred to also as electronic correlation. Møller–Plesset perturbation theory (MPn) and coupled cluster theory (CC) are examples of these post-Hartree–Fock methods. In some cases, particularly for bond breaking processes, the Hartree–Fock method is inadequate and this single-determinant reference function is not a good basis for post-Hartree–Fock methods. It is then necessary to start with a wave function that includes more than one determinant such as multi-configurational self-consistent field (MCSCF) and methods have been developed that use these multi-determinant references for improvements. However, if one uses coupled cluster methods such as CCSDT, CCSDt, CR-CC(2,3), or CC(t;3) then single-bond breaking using the single determinant HF reference is feasible. For an accurate description of double bond breaking, methods such as CCSDTQ, CCSDTq, CCSDtq, CR-CC(2,4), or CC(tq;3,4) also make use of the single determinant HF reference, and do not require one to use multi-reference methods.

Example

Is the bonding situation in disilyne Si_2H_2 the same as in acetylene (C_2H_2)?

A series of *ab initio* studies of Si_2H_2 is an example of how *ab initio* computational chemistry can predict new structures that are subsequently confirmed by experiment. They go back over 20 years, and most of the main conclusions were reached by 1995. The methods used were mostly post-Hartree–Fock, particularly configuration interaction (CI) and coupled cluster (CC). Initially the question was whether disilyne, Si_2H_2 had the same structure as ethyne (acetylene), C_2H_2. In early studies, by Binkley and Lischka and Kohler, it became clear that linear Si_2H_2 was a transition structure between two equivalent trans-bent structures and that the ground state was predicted to be a four-membered ring bent in a 'butterfly' structure with hydrogen atoms bridged between the two silicon atoms. Interest then moved to look at whether structures equivalent to vinylidene ($Si=SiH_2$) existed. This structure is predicted to be a local minimum, i. e. an isomer of Si_2H_2, lying higher in energy than the ground state but below the energy of the trans-bent isomer. Then a new isomer with an unusual structure was predicted by Brenda Colegrove in Henry F. Schaefer, III's group. It requires post-Hartree–Fock methods to obtain a local minimum for this structure. It does not exist on the Hartree–Fock energy hypersurface. The new isomer is a planar structure with one bridging hydrogen atom and one terminal hydrogen atom, cis to the bridging atom. Its energy is above the ground state but below that of the other isomers. Similar results were later obtained for Ge_2H_2. Al_2H_2 and Ga_2H_2 have exactly the same isomers, in spite of having two electrons less than the Group 14 molecules. The only difference is that the four-membered ring ground state is planar and not bent. The cis-mono-bridged and vinylidene-like isomers are present. Experimental work on these molecules is not easy, but matrix isolation spectroscopy of the products of the reaction of hydrogen atoms and silicon and aluminium surfaces has found the ground state ring structures and the cis-mono-bridged structures for Si_2H_2 and Al_2H_2. Theoretical predictions of the vibrational frequencies were crucial in understanding the experimental observations of the spectra of a mixture of compounds. This may appear to be an obscure area of chemistry, but the differences between carbon and silicon chemistry is always a lively question, as are the differences between group 13 and group 14 (mainly the B and C differences). The silicon and germanium compounds were the subject of a Journal of Chemical Education article.

Valence Bond Methods

Valence bond (VB) methods are generally *ab initio* although some semi-empirical versions have been proposed. Current VB approaches are:-

- Generalized valence bond (GVB)
- Modern valence bond theory (MVBT)

Quantum Monte Carlo Methods

A method that avoids making the variational overestimation of HF in the first place is Quantum Monte Carlo (QMC), in its variational, diffusion, and Green's function forms. These methods work with an explicitly correlated wave function and evaluate integrals numerically using a Monte Carlo integration. Such calculations can be very time-consuming. The accuracy of QMC depends strongly on the initial guess of many-body wave-functions and the form of the many-body wave-function. One simple choice is Slater-Jastrow wave-function in which the local correlations are treated with the Jastrow factor.

References

- Griffiths, D. J. (1995). Introduction to Quantum Mechanics. Upper Saddle River, New Jersey: Prentice Hall. ISBN 0-13-124405-1.
- Cropper, William H. (2004), Great Physicists: The Life and Times of Leading Physicists from Galileo to Hawking, Oxford University Press, p. 34, ISBN 978-0-19-517324-6.

- Sakurai, J.J., and Napolitano, J. (1964,2011). Modern quantum mechanics (2nd ed.), Addison Wesley ISBN 978-0-8053-8291-4 .

- Bender, Carl M. and Orszag, Steven A. Advanced Mathematical Methods for Scientists and Engineers. Springer, 1999. ISBN 978-0-387-98931-0

- Verhulst, Ferdinand. Methods and Applications of Singular Perturbations: Boundary Layers and Multiple Timescale Dynamics, Springer, 2005. ISBN 0-387-22966-3.

- Levine, Ira N. (1991). Quantum Chemistry. Englewood Cliffs, New jersey: Prentice Hall. pp. 455–544. ISBN 0-205-12770-3.

- Jensen, Frank (2007). Introduction to Computational Chemistry. Chichester, England: John Wiley and Sons. pp. 80–81. ISBN 0-470-01187-4.

- Cramer, Christopher J. (2002). Essentials of Computational Chemistry. Chichester: John Wiley & Sons, Ltd. pp. 191–232. ISBN 0-471-48552-7.

- Jensen, Frank (2007). Introduction to Computational Chemistry. Chichester, England: John Wiley and Sons. pp. 98–149. ISBN 0-470-01187-4.

Semi-Empirical Quantum Chemistry Method

Semi-empirical quantum chemistry Method is an important concept in computational chemistry. Along with semi-empirical quantum chemistry, this chapter also focuses on extended Hückel method, PM3 and zero differential overlap. The section strategically encompasses and incorporates the major components of semi-empirical quantum chemistry method, providing a complete understanding.

Semi-empirical Quantum Chemistry Method

Semi-empirical quantum chemistry methods are based on the Hartree–Fock formalism, but make many approximations and obtain some parameters from empirical data. They are very important in computational chemistry for treating large molecules where the full Hartree–Fock method without the approximations is too expensive. The use of empirical parameters appears to allow some inclusion of electron correlation effects into the methods.

Within the framework of Hartree–Fock calculations, some pieces of information (such as two-electron integrals) are sometimes approximated or completely omitted. In order to correct for this loss, semi-empirical methods are parametrized, that is their results are fitted by a set of parameters, normally in such a way as to produce results that best agree with experimental data, but sometimes to agree with *ab initio* results.

Type of Simplifications used

Semi-empirical methods follow what are often called empirical methods where the two-electron part of the Hamiltonian is not explicitly included. For π-electron systems, this was the Hückel method proposed by Erich Hückel. For all valence electron systems, the extended Hückel method was proposed by Roald Hoffmann.

Semi-empirical calculations are much faster than their *ab initio* counterparts, mostly due to the use of the zero differential overlap approximation. Their results, however, can be very wrong if the molecule being computed is not similar enough to the molecules in the database used to parametrize the method.

Preferred Application Domains

Semi-empirical calculations have been most successful in the description of organic chemistry, where only a few elements are used extensively and molecules are of moderate size. However, semi-empirical methods were also applied to solids and nanostructures but with different parameterization.

Empirical research is a way of gaining knowledge by means of direct and indirect observation or experience. As with empirical methods, we can distinguish methods that are:

Methods Restricted to π-electrons

These method exist for the calculation of electronically excited states of polyenes, both cyclic and linear. These methods, such as the Pariser–Parr–Pople method (PPP), can provide good estimates of the π-electronic excited states, when parameterized well. Indeed, for many years, the PPP method outperformed ab initio excited state calculations.

Methods Restricted to all Valence Electrons.

These methods can be grouped into several groups:

- Methods such as CNDO/2, INDO and NDDO that were introduced by John Pople. The implementations aimed to fit, not experiment, but ab initio minimum basis set results. These methods are now rarely used but the methodology is often the basis of later methods.

- Methods that are in the MOPAC, AMPAC, and/or SPARTAN computer programs originally from the group of Michael Dewar. These are MINDO, MNDO, AM1, PM3, RM1 , PM6 and SAM1. Here the objective is to use parameters to fit experimental heats of formation, dipole moments, ionization potentials, and geometries.

- Methods whose primary aim is to predict the geometries of coordination compounds, such as Sparkle/AM1, available for lanthanide complexes.

- Methods whose primary aim is to calculate excited states and hence predict electronic spectra. These include ZINDO and SINDO.

Extended Hückel Method

The extended Hückel method is a semiempirical quantum chemistry method, developed by Roald Hoffmann since 1963. It is based on the Hückel method but, while the original Hückel method only considers pi orbitals, the extended method also includes the sigma orbitals.

The extended Hückel method can be used for determining the molecular orbitals, but it is not very successful in determining the structural geometry of an organic molecule. It can however determine the relative energy of different geometrical configurations. It involves calculations of the electronic interactions in a rather simple way for which the electron-electron repulsions are not explicitly included and the total energy is just a sum of terms for each electron in the molecule. The off-diagonal elements of the Hamiltonian matrix are given by an approximation due to Wolfsberg and Helmholz that relates them to the diagonal elements and the overlap matrix element.

$$H_{ij} = KS_{ij} \frac{H_{ii} + H_{jj}}{2}$$

K is the Wolfsberg-Helmholtz constant, and is usually given a value of 1.75. In the extended Hückel method, only valence electrons are considered; the core electron energies and functions are supposed to be more or less constant between atoms of the same type. The method uses a series of parametrized energies calculated from atomic ionisation potentials or theoretical methods to fill the diagonal of the Fock matrix. After filling the non-diagonal elements and diagonalizing the resulting Fock matrix, the energies (eigenvalues) and wavefunctions (eigenvectors) of the valence orbitals are found.

It is common in many theoretical studies to use the extended Hückel molecular orbitals as a preliminary step to determining the molecular orbitals by a more sophisticated method such as the CNDO/2 method and ab initio quantum chemistry methods. Since the extended Hückel basis set is fixed, the monoparticle calculated wavefunctions must be projected to the basis set where the accurate calculation is to be done. One usually does this by adjusting the orbitals in the new basis to the old ones by least squares method. As only valence electron wavefunctions are found by this method, one must fill the core electron functions by orthonormalizing the rest of the basis set with the calculated orbitals and then selecting the ones with less energy. This leads to the determination of more accurate structures and electronic properties, or in the case of ab initio methods, to somewhat faster convergence.

The method was first used by Roald Hoffmann who developed, with Robert Burns Woodward, rules for elucidating reaction mechanisms (the Woodward–Hoffmann rules). He used pictures of the molecular orbitals from extended Hückel theory to work out the orbital interactions in these cycloaddition reactions.

A closely similar method was used earlier by Hoffmann and William Lipscomb for studies of boron hydrides. The off-diagonal Hamiltonian matrix elements were given as proportional to the overlap integral.

$$H_{ij} = KS_{ij}$$

This simplification of the Wolfsberg and Helmholz approximation is reasonable for boron hydrides as the diagonal elements are reasonably similar due to the small difference in electronegativity between boron and hydrogen.

The method works poorly for molecules that contain atoms of very different electronegativity. To overcome this weakness, several groups have suggested iterative schemes that depend on the atomic charge. One such method, that is still widely used in inorganic and organometallic chemistry is the Fenske-Hall method.

A recent program for the *extended Hückel method* is YAeHMOP which stands for "yet another extended Hückel molecular orbital package".

PM3 (Chemistry)

Ball-and-stick model of the aplysin molecule, C15H19BrO. Colour code: Carbon, C: black Hydrogen, H: white Bromine, Br: red-brown Oxygen, O: red Structure calculated with Spartan Student 4.1, using the PM3 semi-empirical method.

PM3, or Parameterized Model number 3, is a semi-empirical method for the quantum calculation of molecular electronic structure in computational chemistry. It is based on the Neglect of Differential Diatomic Overlap integral approximation.

The PM3 method uses the same formalism and equations as the AM1 method. The only differences are: 1) PM3 uses two Gaussian functions for the core repulsion function, instead of the variable number used by AM1 (which uses between one and four Gaussians per element); 2) the numerical values of the parameters are different. The other differences lie in the philosophy and methodology used during the parameterization: whereas AM1 takes some of the parameter values from spectroscopical measurements, PM3 treats them as optimizable values.

The method was developed by J. J. P. Stewart and first published in 1989. It is implemented in the MOPAC program (of which the older versions are public domain), along with the related RM1, AM1, MNDO and MINDO methods, and in several other programs such as Gaussian, CP2K, GAMESS (US), GAMESS (UK), PC GAMESS, Chem3D, AMPAC, ArgusLab, BOSS, and SPARTAN.

The original PM3 publication included parameters for the following elements: H, C, N, O, F, Al, Si, P, S, Cl, Br, and I.

The PM3 implementation in the SPARTAN program includes PM3tm with additional extensions for transition metals supporting calculations on Ca, Ti, V, Cr, Mn, Fe, Co, Ni, Cu, Zn, Zr, Mo, Tc, Ru, Rh, Pd, Hf, Ta, W, Re, Os, Ir, Pt, and Gd. Many other elements, mostly metals, have been parameterized in subsequent work.

A model for the PM3 calculation of lanthanide complexes, called Sparkle/PM3, was also introduced.

Zero Differential Overlap

Zero differential overlap is an approximation in computational molecular orbital theory that is the central technique of semi-empirical methods in quantum chemistry. When computers were first used to calculate bonding in molecules, it was possible to only calculate diatomic molecules. As computers advanced, it became possible to study larger molecules, but the use of this approximation has always allowed the study of even larger molecules. Currently semi-empirical methods can be applied to molecules as large as whole proteins. The approximation involves ignoring certain integrals, usually two-electron repulsion integrals. If the number of orbitals used in the calculation is N, the number of two-electron repulsion integrals scales as N^4. After the approximation is applied the number of such integrals scales as N^2, a much smaller number, simplifying the calculation.

Details of Approximation

If the molecular orbitals Φi are expanded in terms of N basis functions, χ_μ^A as:-

$$\Phi_i = \sum_{\mu=1}^{N} C_{i\mu} \, \chi_\mu^A$$

where A is the atom the basis function is centred on, and $C_{i\mu}$ are coefficients, the two-electron repulsion integrals are then defined as:-

$$\langle \mu v \mid \lambda \sigma \rangle = \iint \chi_\mu^A(1) \chi_v^B(1) \frac{1}{r_{12}} \chi_\lambda^C(2) \chi_\sigma^D(2) d\tau_1 d\tau_2$$

The zero differential overlap approximation ignores integrals that contain the product $\chi_\mu^A(1)\chi_v^B(1)$ where μ is not equal to v. This leads to:-

$$\langle \mu v \mid \lambda \sigma \rangle = \delta_{\mu v} \delta_{\lambda \sigma} \langle \mu\mu \mid \lambda\lambda \rangle$$

where $\delta_{\mu v} = \begin{cases} 0 & \mu \neq v \\ 1 & \mu = v \end{cases}$

The total number of such integrals is reduced to $N(N + 1) / 2$ (approximately $N^2 / 2$) from $[N(N + 1) / 2][N(N + 1) / 2 + 1] / 2$ (approximately $N^4 / 8$), all of which are included in ab initio Hartree–Fock and post-Hartree–Fock calculations.

Scope of Approximation in Semi-empirical Methods

Methods such as the Pariser–Parr–Pople method (PPP) and CNDO/2 use the zero differential overlap approximation completely. Methods based on the intermediate neglect of differential overlap, such as INDO, MINDO, ZINDO and SINDO do not apply it when $A = B = C = D$, i.e. when all four basis functions are on the same atom. Methods that use the neglect of diatomic differential overlap, such as MNDO, PM3 and AM1, also do not apply it when $A = B$ and $C = D$, i.e. when the basis functions for the first electron are on the same atom and the basis functions for the second electron are the same atom.

It is possible to partly justify this approximation, but generally it is used because it works reasonably well when the integrals that remain – $\langle \mu\mu \,|\, \lambda\lambda \rangle$ – are parameterised.

Understanding Diatomic Molecule

The molecules formed by only two atoms are referred to as diatomic molecules and diatomic molecules can be composed of the same atom in case of oxygen (O2) or of two different molecules like nitric oxide (NO). This chapter explores the subject of diatomic molecules by imputing to Hund's cases and the symmetry of diatomic molecules.

Diatomic Molecule

Diatomic molecules are molecules composed of only two atoms, of the same or different chemical elements. The prefix *di-* is of Greek origin, meaning "two". If a diatomic molecule consists of two atoms of the same element, such as hydrogen (H_2) or oxygen (O_2), then it is said to be homonuclear. Otherwise, if a diatomic molecule consists of two different atoms, such as carbon monoxide (CO) or nitric oxide (NO), the molecule is said to be heteronuclear.

A space-filling model of the diatomic molecule dinitrogen, N_2

A periodic table showing the elements that exist as homonuclear diatomic molecules under typical laboratory conditions.

The only chemical elements that form stable homonuclear diatomic molecules at standard temperature and pressure (STP) (or typical laboratory conditions of 1 bar and 25 °C) are the gases hydrogen (H_2), nitrogen (N_2), oxygen (O_2), fluorine (F_2), and chlorine (Cl_2).

The noble gases (helium, neon, argon, krypton, xenon, and radon) are also gases at STP, but they are monatomic. The homonuclear diatomic gases and noble gases together are called "elemental gases" or "molecular gases", to distinguish them from other gases that are chemical compounds.

At slightly elevated temperatures, the halogens bromine (Br_2) and iodine (I_2) also form diatomic gases. All halogens have been observed as diatomic molecules, except for astatine, which is uncertain.

Group →	1	2	3	4	5	6	7	8	9	10	11	12	13	14	15	16	17	18
Period 1	1 H																	2 He
2	3 Li	4 Be											5 B	6 C	7 N	8 O	9 F	10 Ne
3	11 Na	12 Mg											13 Al	14 Si	15 P	16 S	17 Cl	18 Ar
4	19 K	20 Ca	21 Sc	22 Ti	23 V	24 Cr	25 Mn	26 Fe	27 Co	28 Ni	29 Cu	30 Zn	31 Ga	32 Ge	33 As	34 Se	35 Br	36 Kr
5	37 Rb	38 Sr	39 Y	40 Zr	41 Nb	42 Mo	43 Tc	44 Ru	45 Rh	46 Pd	47 Ag	48 Cd	49 In	50 Sn	51 Sb	52 Te	53 I	54 Xe
6	55 Cs	56 Ba		72 Hf	73 Ta	74 W	75 Re	76 Os	77 Ir	78 Pt	79 Au	80 Hg	81 Tl	82 Pb	83 Bi	84 Po	85 At	86 Rn
7	87 Fr	88 Ra		104 Rf	105 Db	106 Sg	107 Bh	108 Hs	109 Mt	110 Ds	111 Rg	112 Cn	113 Uut	114 Fl	115 Uup	116 Lv	117 Uus	118 Uuo

Lanthanides	57 La	58 Ce	59 Pr	60 Nd	61 Pm	62 Sm	63 Eu	64 Gd	65 Tb	66 Dy	67 Ho	68 Er	69 Tm	70 Yb	71 Lu
Actinides	89 Ac	90 Th	91 Pa	92 U	93 Np	94 Pu	95 Am	96 Cm	97 Bk	98 Cf	99 Es	100 Fm	101 Md	102 No	103 Lr

Other elements form diatomic molecules when evaporated, but these diatomic species repolymerize when cooled. Heating ("cracking") elemental phosphorus gives diphosphorus, P_2. Sulfur vapor is mostly disulfur (S_2). Dilithium (Li_2) is known in the gas phase. Ditungsten (W_2) and dimolybdenum (Mo_2) form with sextuple bonds in the gas phase. The bond in a homonuclear diatomic molecule is non-polar.

Heteronuclear Molecules

All other diatomic molecules are chemical compounds of two different elements. Many elements can combine to form heteronuclear diatomic molecules, depending on temperature and pressure.

Common examples include the gases carbon monoxide (CO), nitric oxide (NO), and hydrogen chloride (HCl).

Many 1:1 binary compounds are not normally considered diatomic because they are polymeric at room temperature, but they form diatomic molecules when evaporated, for example gaseous MgO, SiO, and many others.

Occurrence

Hundreds of diatomic molecules have been identified in the environment of the Earth, in the laboratory, and in interstellar space. About 99% of the Earth's atmosphere is composed of two species of diatomic molecules: nitrogen (78%) and oxygen (21%). The natural abundance of hydrogen (H_2) in the Earth's atmosphere is only of the order of parts per million, but H_2 is the most abundant diatomic molecule in the universe. The interstellar medium is, indeed, dominated by hydrogen atoms.

Molecular Geometry

Diatomic molecules cannot have any geometry but linear, as any two points always lie in a straight line. This is the simplest spatial arrangement of atoms.

Historical Significance

Diatomic elements played an important role in the elucidation of the concepts of element, atom, and molecule in the 19th century, because some of the most common elements, such as hydrogen, oxygen, and nitrogen, occur as diatomic molecules. John Dalton's original atomic hypothesis assumed that all elements were monatomic and that the atoms in compounds would normally have the simplest atomic ratios with respect to one another. For example, Dalton assumed water's formula to be HO, giving the atomic weight of oxygen as eight times that of hydrogen, instead of the modern value of about 16. As a consequence, confusion existed regarding atomic weights and molecular formulas for about half a century.

As early as 1805, Gay-Lussac and von Humboldt showed that water is formed of two volumes of hydrogen and one volume of oxygen, and by 1811 Amedeo Avogadro had arrived at the correct interpretation of water's composition, based on what is now called Avogadro's law and the assumption of diatomic elemental molecules. However, these results were mostly ignored until 1860, partly due to the belief that atoms of one element would have no chemical affinity toward atoms of the same element, and partly due to apparent exceptions to Avogadro's law that were not explained until later in terms of dissociating molecules.

At the 1860 Karlsruhe Congress on atomic weights, Cannizzaro resurrected Avogadro's ideas and used them to produce a consistent table of atomic weights, which mostly agree with modern values. These weights were an important prerequisite for the discovery of the periodic law by Dmitri Mendeleev and Lothar Meyer.

Excited Electronic States

Diatomic molecules are normally in their lowest or ground state, which conventionally is also known as the X state. When a gas of diatomic molecules is bombarded by energetic electrons, some of the molecules may be excited to higher electronic states, as occurs, for example, in the natural aurora; high-altitude nuclear explosions; and rocket-borne electron gun experiments. Such excitation can also occur when the gas absorbs light or other electromagnetic radiation. The excited states are unstable and naturally relax back to the ground state. Over various short time scales after the excitation (typically a fraction of a second, or sometimes longer than a second if the excited state is metastable), transitions occur from higher to lower electronic states and ultimately to the ground state, and in each transition results a photon is emitted. This emission is known as fluorescence. Successively higher electronic states are conventionally named

A, B, C, etc. (but this convention is not always followed, and sometimes lower case letters and alphabetically out-of-sequence letters are used, as in the example given below). The excitation energy must be greater than or equal to the energy of the electronic state in order for the excitation to occur.

In quantum theory, an electronic state of a diatomic molecule is represented by

$$^{2S+1} \Lambda(v)$$

where S is the total electronic spin quantum number, Λ is the total electronic angular momentum quantum number along the internuclear axis, and v is the vibrational quantum number. Λ takes on values 0, 1, 2, ..., which are represented by the electronic state symbols Σ, Π, Δ,.... For example, the following table lists the common electronic states (without vibrational quantum numbers) along with the energy of the lowest vibrational level ($v = 0$) of diatomic nitrogen (N_2), the most abundant gas in the Earth's atmosphere. In the table, the subscripts and superscripts after Λ give additional quantum mechanical details about the electronic state.

State	Energy (T_0, cm^{-1}) See note below
$X^1\Sigma_g^+$	0.0
$A^3\Sigma_u^+$	49754.8
$B^3\Pi_g$	59306.8
$W^3\Delta_u$	59380.2
$B'^3\Sigma_u^-$	65851.3
$a'^1\Sigma_u^-$	67739.3
$a^1\Pi_g$	68951.2
$w^1\Delta_u$	71698.4

Note: The "energy" units in the above table are actually the reciprocal of the wavelength of a photon emitted in a transition to the lowest energy state. The actual energy can be found by multiplying the given statistic by the product of c (the speed of light) and h (Planck's constant), i.e., about 1.99×10^{-25} Joule metres, and then multiplying by a further factor of 100 to convert from cm^{-1} to m^{-1}.

The aforementioned fluorescence occurs in distinct regions of the electromagnetic spectrum, called "emission bands": each band corresponds to a particular transition from a higher electronic state and vibrational level to a lower electronic state and vibrational level (typically, many vibrational levels are involved in an excited gas of diatomic molecules). For example, N_2 A - X X emission bands (a.k.a. Vegard-Kaplan bands) are present in the spectral range from 0.14 to 1.45 µm (micrometres). A given band can be spread out over several nanometers in electromagnetic wavelength space, owing to the various transitions that occur in the molecule's rotational quantum number, J . These are classified into distinct sub-band branches, depending on the change in J . The R branch corresponds to $\Delta J = +1$, the P branch to $\Delta J = -1$, and the Q branch to $\Delta J = 0$. Bands are spread out even further by the limited spectral resolution of the spectrometer that is used to measure the spectrum. The spectral resolution depends on the instrument's point spread function.

Energy Levels

The molecular term symbol is a shorthand expression of the angular momenta that characterize the electronic quantum states of a diatomic molecule, which are eigenstates of the electronic molecular Hamiltonian. It is also convenient, and common, to represent a diatomic molecule as two point masses connected by a massless spring. The energies involved in the various motions of the molecule can then be broken down into three categories: the translational, rotational, and vibrational energies.

Translational Energies

The translational energy of the molecule is given by the kinetic energy expression:

$$E_{trans} = \frac{1}{2}mv^2$$

where m is the mass of the molecule and v is its velocity.

Rotational Energies

Classically, the kinetic energy of rotation is

$$E_{rot} = \frac{L^2}{2I}$$

where

L is the angular momentum

I is the moment of inertia of the molecule

For microscopic, atomic-level systems like a molecule, angular momentum can only have specific discrete values given by

$$L^2 = l(l+1)\hbar^2$$

where l is a non-negative integer and $\hbar$ is the reduced Planck constant.

Also, for a diatomic molecule the moment of inertia is

$$I = \mu r_0^2$$

where

μ is the reduced mass of the molecule and

r_0 is the average distance between the centers of the two atoms in the molecule.

So, substituting the angular momentum and moment of inertia into E_{rot}, the rotational energy levels of a diatomic molecule are given by:

$$E_{rot} = \frac{l(l+1)\hbar^2}{2\mu r_0^2} \quad l = 0, 1, 2, ...$$

Vibrational Energies

Another type of motion of a diatomic molecule is for each atom to oscillate—or vibrate—along the line connecting the two atoms. The vibrational energy is approximately that of a quantum harmonic oscillator:

$$E_{vib} = \left(n + \frac{1}{2} \right) \hbar \omega \quad n = 0, 1, 2, ...$$

where

n is an integer

$\hbar$ is the reduced Planck constant and

ω is the angular frequency of the vibration.

Comparison Between Rotational and Vibrational Energy Spacings

The spacing, and the energy of a typical spectroscopic transition, between vibrational energy levels is about 100 times greater than that of a typical transition between rotational energy levels.

Hund's Cases

The good quantum numbers for a diatomic molecule, as well as good approximations of rotational energy levels, can be obtained by modeling the molecule using Hund's cases.

Hund's Cases

In rotational-vibrational and electronic spectroscopy of diatomic molecules, Hund's coupling cases are idealized cases where specific terms appearing in the molecular Hamiltonian and involving couplings between angular momenta are assumed to dominate over all other terms. There are five cases, traditionally notated with the letters (a) through (e). Most diatomic molecules are somewhere between the idealized cases (a) and (b).

Angular Momenta

To describe the Hund's coupling cases, we use the following angular momenta:

- $\mathbf{L}$, the electronic orbital angular momentum
- $\mathbf{S}$, the electronic spin angular momentum
- $\mathbf{J}_a = \mathbf{L} + \mathbf{S}$, the total electronic angular momentum
- $\mathbf{J}$, the total angular momentum of the system
- $\mathbf{N} = \mathbf{J} - \mathbf{S}$, the total angular momentum minus the electron spin
- $\mathbf{R} = \mathbf{N} - \mathbf{L}$, the rotational angular momentum of the nuclei

Choosing the Applicable Hund's Case

Hund's coupling cases are idealizations. The appropriate case for a given situation can be found by comparing three strengths: the electrostatic coupling of $\mathbf{L}$ to the internuclear axis, the spin-orbit coupling, and the rotational coupling of $\mathbf{L}$ and $\mathbf{S}$ to the total angular momentum $\mathbf{J}$.

Hund's case	Electrostatic	Spin-orbit	Rotational
(a)	strong	intermediate	weak
(b)	strong	weak	intermediate
(c)	intermediate	strong	weak
(d)	intermediate	weak	strong
(e)	weak	intermediate	strong
		strong	intermediate

The last two rows are degenerate because they have the same good quantum numbers.

Case (a)

In case (a), $\mathbf{L}$ is electrostatically coupled to the internuclear axis, and $\mathbf{S}$ is coupled to $\mathbf{L}$ by spin-orbit coupling. Then both $\mathbf{L}$ and $\mathbf{S}$ have well-defined axial components Λ and Σ , respectively. $\dot{\mathbf{U}}$ defines a vector of magnitude $\Omega = \Lambda + \Sigma$ pointing along the internuclear axis. Combined with the rotational angular momentum of the nuclei $\mathbf{R}$, we have $J = \Omega + R$. In this case, the precession of $\mathbf{L}$ and $\mathbf{S}$ around the nuclear axis is assumed to be much faster than the nutation of $\dot{\mathbf{U}}$ and $\mathbf{R}$ around $\mathbf{J}$.

The good quantum numbers in case (a) are Λ , S , Σ , J and Ω . We express the rotational energy operator as $H_{rot} = B\mathbf{R}^2 = B(\mathbf{J} - \mathbf{L} - \mathbf{S})^2$, where B is a rotational constant.

There are, ideally, $2S+1$ fine-structure states, each with rotational levels having relative energies starting with $J = \Omega$.

Case (b)

In case (b), the spin-orbit coupling is weak or non-existent (in the case $\Lambda = 0$). In this case, we take $\mathbf{N} = \ddot{e} + \mathbf{R}$ and $\mathbf{J} = \mathbf{N} + \mathbf{S}$ and assume $\mathbf{L}$ precesses quickly around the internuclear axis.

The good quantum numbers in case (b) are Λ, N, S, and J. We express the rotational energy operator as $H_{rot} = B\mathbf{R}^2 = B(\mathbf{N} - \mathbf{L})^2$, where B is a rotational constant. The rotational levels therefore have relative energies $BN(N+1)$ starting with $N = \Lambda$.

Case (c)

In case (c), the spin-orbit coupling is stronger than the coupling to the internuclear axis, and Λ and Σ from case (a) cannot be defined. Instead $\mathbf{L}$ and $\mathbf{S}$ combine to form $\mathbf{J}_a$, which has a projection along the internuclear axis of magnitude Ω. Then $J = \Omega + R$, as in case (a).

The good quantum numbers in case (c) are J_a, J, and Ω.

Case (d)

In case (d), the rotational coupling between $\mathbf{L}$ and $\mathbf{R}$ is much stronger than the electrostatic coupling of $\mathbf{L}$ to the internuclear axis. Thus we form $\mathbf{N}$ by coupling $\mathbf{L}$ and $\mathbf{R}$ and the form by coupling $\mathbf{J}$ and $\mathbf{S}$.

The good quantum numbers in case (d) are L, R, N, S, and J. Because R is a good quantum number, the rotational energy is simply $H_{rot} = B\mathbf{R}^2 = BR(R+1)$.

Case (e)

In case (e), we first form $\mathbf{J}_a$ and then form $\mathbf{J}$ by coupling $\mathbf{J}_a$ and $\mathbf{R}$. This case is rare but has been observed.

The good quantum numbers in case (e) are J_a, R, and J. Because R is once again a good quantum number, the rotational energy is $H_{rot} = B\mathbf{R}^2 = BR(R+1)$.

Symmetry of Diatomic Molecules

Molecular symmetry in physics and chemistry describes the symmetry present in molecules and the classification of molecules according to their symmetry. Molecular symmetry is a fundamental concept in the application of Quantum Mechanics in physics and chemistry, for example it can be used to predict or explain many of a molecule's properties, such as its dipole moment and its allowed spectroscopic transitions (based on selection rules), without doing the exact rigorous calculations (which, in some cases, may not even be possible). Group theory is the predominant framework for analyzing molecular symmetry.

Among all the molecular symmetries, diatomic molecules show some distinct features and they are relatively easier to analyze.

Symmetry and Group Theory

The physical laws governing a system is generally written as a relation (equations, differential equations, integral equations etc.). An operation on the ingredients of this relation, which keeps the form of the relations invariant is called a symmetry transformation or a symmetry of the system.

- These symmetry operations can involve external or internal co-ordinates; giving rise to geometrical or internal symmetries.

- These symmetry operations can be global or local; giving rise to global or gauge symmetries.

- These symmetry operations can be discrete or continuous.

Symmetry is a fundamentally important concept in quantum mechanics. It can predict conserved quantities and provide quantum numbers. It can predict degeneracies of eigenstates and gives insights about the matrix elements of the Hamiltonian without calculating them. Rather than looking into individual symmetries, it is sometimes more convenient to look into the general relations between the symmetries. It turns out that Group theory is the most efficient way of doing this.

Groups

A *group* is a mathematical structure (usually denoted in the form $(G,*)$) consisting of a set G and a binary operation '*' (sometimes loosely called 'multiplication'), satisfying the following properties:

1. closure: For every pair of elements $x, y \in G$, the *product* $x * y \in G$.

2. associativity: For every x and y and z in G, both $(x*y)*z$ and $x*(y*z)$ result with the same element in G (in symbols, $(x * y) * z = x * (y * z) \forall x, y, z \in G$).

3. existence of identity: There must be an element (say e) in G such that product any element of G with e make no change to the element (in symbols, $x * e = e * x = x; \forall x \in G$).

4. existence of inverse: For each element (x) in G, there must be an element y in G such that product of x and y is the identity element e (in symbols, for each $x \in G$ $\exists\, y \in G$ such that $x*y = y*x = e$).

- In addition to the above four, if it so happens that $\forall x, y \in G$, $\exists\, y \in G$, i.e., the operation in commutative, then the group is called an Abelian Group. Otherwise it is called a Non-Abelian Group.

Groups, Symmetry and Conservation

The set of all symmetry transformations of a Hamiltonian has the structure of a group, with group multiplication equivalent to applying the transformations one after the other. The group elements can be represented as matrices, so that the group operation becomes the ordinary matrix multiplication. In quantum mechanics, the evolution of an arbitrary superposition of states are given by unitary operators, so each of the elements of the symmetry groups are unitary operators. Now any unitary operator can be expressed as the exponential of some Hermitian operator . So, the corresponding Hermitian operators are the 'generators' of the symmetry group. These unitary transformations act on the Hamiltonian operator in some Hilbert space in a way that the Hamiltonian remains invariant under the transformations. In other words, the symmetry operators commute with the Hamiltonian. If U represents the unitary symmetry operator and acts on the Hamiltonian H , then;

$$[H,U]=0$$

$$H' = U^{\dagger}HU = H$$
$$\Rightarrow HU = UH$$
$$\Rightarrow [H,U]= 0;\forall U \in G$$

These operators have the above-mentioned properties of a group:

- The symmetry operations are closed under multiplication.

- Application of symmetry transformations are associative.

- There is always a trivial transformation, where nothing is done to the original co-ordinates. This is the identity element of the group.

- And as long as an inverse transformation exists, it is a symmetry transformation, i.e. it leaves the Hamiltonian invariant. Thus the inverse is part of this set.

So, by the symmetry of a system, we mean a set of operators, each of which commutes with the Hamiltonian, and they form a symmetry group. This group may be Abelian or Non-Abelian. Depending upon which one it is, the properties of the system changes (for example, if the group is Abelian, there would be no degeneracy). Corresponding to every different kind of symmetry in a system, we can find a symmetry group associated with it.

It follows that the generator T of the symmetry group also commutes with the Hamiltonian. Now, it follows that:

The observable corresponding to the generator Hermitian matrix, is conserved.

The derivative of the expectation value of the operator T can be written as:

$$\frac{d\langle T\rangle}{dt} = \frac{d\langle \Psi\,|\,T\,|\,\Psi\rangle}{dt} = \left\langle \frac{\partial\Psi}{\partial t}\,|\,T\,|\,\Psi\right\rangle + \left\langle \Psi\,|\,\frac{\partial T}{\partial t}\,|\,\Psi\right\rangle + \left\langle \Psi\,|\,T\,|\,\frac{\partial\Psi}{\partial t}\right\rangle$$

Now,

$$i\hbar\frac{\partial|\Psi\rangle}{\partial t} = H|\Psi\rangle$$

So,

$$\frac{d\langle T\rangle}{dt} = -\frac{1}{i\hbar}\langle \Psi\,|\,HT\,|\,\Psi\rangle + \frac{1}{i\hbar}\langle \Psi\,|\,TH\,|\,\Psi\rangle + \left\langle \frac{\partial T}{\partial t}\right\rangle$$

as H is also Hermitian. So we have,

$$\frac{d\langle T\rangle}{dt} = \frac{1}{i\hbar}\langle[H,T]\rangle + \left\langle \frac{\partial T}{\partial t}\right\rangle \quad \text{Now, } [H,T]=0$$

as stated above, and if the operator T does not have any explicit time-dependence;

$$\frac{d\langle T\rangle}{dt} = 0 \Rightarrow \langle T\rangle \text{ is a constant, independent of what the state } |\Psi\rangle \text{ may be.}$$

So the observable corresponding to the operator T, is conserved.

Some specific examples can be systems having rotational, translational invariance etc.. For a rotationally invariant system, The symmetry group of the Hamiltonian is the general rotation group. Now, if (say) the system is invariant about any rotation about Z-axis (i.e., the system has axial symmetry), then the symmetry group of the Hamiltonian is the group of rotation about the symmetry axis. Now, this group is generated by the Z-component of the orbital angular momentum, L_z (general group element $R(\alpha) = e^{\frac{-i\alpha L_z}{\hbar}}$). Thus, L_z commutes with H for this system and Z-component of the angular momentum is conserved. Similarly, translation symmetry gives rise to conservation of linear momentum, inversion symmetry gives rise to parity conservation and so on.

Geometrical Symmetries

Symmetry Operations, Symmetry Elements and Point Group

All the molecules (or rather the molecular models) possess certain geometrical symmetries. The application of the corresponding symmetry operation produces a spatial orientation of the molecule which is indistinguishable from the previous one. There are predominantly five main types of symmetry operations: identity, rotation, reflection, inversion and improper rotation or rotation-reflection. Corresponding to each symmetry operation there is a corresponding symmetry element, with respect to which the symmetry operation is applied. Common to all symmetry operations is that the geometrical center of a molecule does not change its position, all symmetry elements must intersect in this center. Thus, these symmetry operations make a special kind of group, named point groups. On the contrary, there exists another kind of group important in crystallography, where translation in 3-D also needs to be taken care of. They are known as space groups.

Basic Symmetry Operations

The five basic symmetry operations mentioned above are:

1. Identity Operation E (from the German 'Einheit' meaning unity):The identity operation leaves the molecule unchanged. It forms the identity element in the symmetry group. Though its inclusion seems to be trivial, it is important also because even for the most asymmetric molecule, this symmetry is present. The corresponding symmetry element is the entire molecule itself.

2. Inversion, i : This operation inverts the molecule about its center of inversion (if it has any). The center of inversion is the symmetry element in this case. There may or may not be an atom at this center. A molecule may or may not have a center of inversion. For example: the benzene molecule, a cube, and spheres do have a center of inversion, whereas a tetrahedron does not.

3. Reflection σ: The reflection operation produces a mirror image geometry of the molecule about a certain plane. The mirror plane bisects the molecule and must include its center of geometry. The plane of symmetry is the symmetry element in this case. A symmetry plane parallel with the principal axis (defined below) is dubbed vertical (σ_v) and one perpendicular to it horizontal (σ_h). A third type of symmetry plane exists: If a vertical symmetry plane additionally bisects the angle between two 2-fold rotation axes perpendicular to the principal axis, the plane is dubbed dihedral (σ_d).

4. n-Fold Rotation c_n : The n-fold rotation operation about a n-fold axis of symmetry produces molecular orientations indistinguishable from the initial for each rotation of $\frac{360^0}{n}$ (clockwise and counter-clockwise).It is denoted by c_n . The axis of symmetry is the symmetry element in this case. A molecule can have more than one symmetry axis; the one with the highest n is called the principal axis, and by convention is assigned the z-axis in a Cartesian coordinate system.

5. *n*-Fold Rotation-Reflection or improper rotation S_n : The n-fold improper rotation operation about an n-fold axis of improper rotation is composed of two successive geometry transformations: first, a rotation through $\frac{360°}{n}$ about the axis of that rotation, and second, reflection through a plane perpendicular (and through the molecular center of geometry) to that axis. This axis is the symmetry element in this case. It is abbreviated S_n.

It is to be noted that all other symmetry present in a specific molecule are a combination of these 5 operations.

Schoenflies Notation

The Schoenflies (or Schönflies) notation, named after the German mathematician Arthur Moritz Schoenflies, is one of two conventions commonly used to describe point groups. This notation is used in spectroscopy and is sufficient for the classification of symmetry groups of a molecule. Here onwards, Schoenflies notation will be used to specify a molecular point group. In three dimensions, there are an infinite number of point groups, but all of them can be classified by several families.

Groups with $n = \infty$ are called limit groups or Curie groups. The two symmetry groups of the most importance in the context of diatomic molecules belong to this class. The complete description of these groups require Lie algebra.

Symmetry Groups in Diatomic Molecules

There are typically two symmetry groups associated with diatomic molecules: $C_{\infty v}$ and $D_{\infty h}$, both of them belonging to the Curie groups.

- $C_{\infty v}$:

The simplest axial symmetry group is the group $C_{\infty v}$, which contains which contains rotations $C(\phi)$ through any angle ϕ about the axis of symmetry; this is called the *two-dimensional rotation group*. It may be regarded as the limiting case of the groups C_{nv} as $n \to \infty$.

In the context of diatomic molecules, *every* diatomic molecule is symmetric about reflections σ_v through the planes passing through the inter-nuclear axis (or the vertical axis, that is reason of the subscript '*v*').In the group $C_{\infty v}$ all planes of symmetry are equivalent, so that all reflections σ_v form a single class with a continuous series of elements; the axis of symmetry is bilateral, so that there is a continuous series of classes, each containing two elements $C(\pm\phi)$. Note that this group is *Non-abelian* and there exists ∞ number of irreps of the group.The character table of this group is as follows:

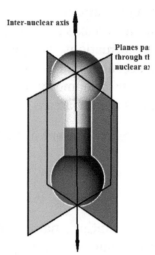

Axial symmetry in diatomic molecules, giving rise to symmetry group $C_{\infty v}$

	E	$2c_\infty$	...	$\infty \sigma_y$	linear functions, rotations	quadratic
$A_1 = \Sigma^+$	1	1	...	1	z	x^2+y^2, z^2
$A_2 = \Sigma^-$	1	1	...	-1	R_z	
$E_1 = \Pi$	2	$2\cos(\phi)$	...	0	(x, y) (R_x, R_y)	x^2+y^2, z^2
$E_2 = \Delta$	2	$2\cos(2\phi)$	...	0		(x^2-y^2, xy)
$E_3 = \Phi$	2	$2\cos(3\phi)$	....	0		
...	...	...	...			

$D_{\infty h}$:

In addition to axial reflection symmetry, homonuclear diatomic molecules are symmetric about inversion or reflection through any axis in the plane passing through the point of symmetry and perpendicular to the inter-nuclear axis.

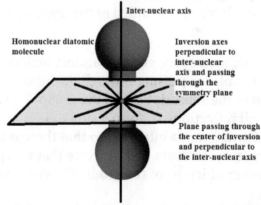

Inversion symmetry in homonuclear diatomic molecules, giving rise to symmetry group

This gives rise to the group $D_{\infty h}$ (a dihedral group) as the other symmetry group (in addition to $C_{\infty v}$). The classes of the group $D_{\infty h}$ can be obtained from those of the group $C_{\infty v}$ by the relation between the two groups: $D_{\infty h} = C_{\infty v} \times C_i$. Like $C_{\infty v}$, $D_{\infty h}$ is *non-Abelian* and there exists ∞ number of irreps of the group. The character table of this group is as follows:

	E	$2c_{\infty}$	...		i	$2S_{\infty}$	...	$\infty c_2'$	linear functions, rotations	quadratic
$A_{1g}=\Sigma^+_g$	1	1	...	1	1	1	...		z	x^2+y^2, z^2
$A_{2g}=\Sigma^-_g$	1	1	...	-1	1	1	...		R_z	
$E_{1g}=\Pi_g$	2	$2\cos(\phi)$	...	0	2	$-2\cos(\phi)$	...		(x, y) (R_x, R_y)	x^2+y^2, z^2
$E_{2g}=\Delta_g$	2	$2\cos(2\phi)$	...	0	2	$2\cos(2\phi)$	...			(x^2-y^2, xy)
$E_{3g}=\Phi_g$	2	$2\cos(3\phi)$	...	0	2	$-2\cos(3\phi)$	...			
...	...	...	...	...	...	...	...			
$A_{1u}=\Sigma^+_u$	1	1	...	1	-1	-1	...		z	
$A_{2u}=\Sigma^-_u$	1	1	...	-1	-1	-1	...			
$E_{1u}=\Pi_u$	2	$2\cos(\phi)$	...	0	-2	$2\cos(\phi)$	...		(x, y)	
$E_{2u}=\Delta_u$	2	$2\cos(2\phi)$	...	0	-2	$-2\cos(2\phi)$	...			
$E_{3u}=\Phi_u$	2		...	0	-2	$2\cos(3\phi)$	...			
...	...	...	...	...	...	...	...			

Summary Examples

Point group	Symmetry operations or group operations	Symmetry elements or group elements	Simple description of typical geometry	Group order	Number of classes and irreducible representations (irreps)	Example
$C_{\infty v}$	$E, c^{\phi}_{\infty}, \sigma_v$	$E, 2c_{\infty}, \infty\sigma_v$	linear	∞	∞	Hydrogen fluoride

$D_{\infty h}$	E, c_∞^ϕ, σ_h, i, c_2'	$S_\infty, E, 2c_\infty,$ $\infty\sigma_h, \infty c_2'$	linear with inversion center	∞	∞	
						oxygen

Complete Set of Commuting Operators

Unlike a single atom, the Hamiltonian of a diatomic molecule doesn't commute with L^2. So the quantum number l is no longer a good quantum number. The internuclear axis chooses a specific direction in space and the potential is no longer spherically symmetric. Instead, L_z and J_z commutes with the Hamiltonian H (taking the arbitrary internuclear axis as the Z axis). But L_x, L_y do not commute with H due to the fact that the electronic Hamiltonian of a diatomic molecule is invariant under rotations about the internuclear line (the Z axis), but not under rotations about the X or Y axes. Again, S^2 and S_z act on a different Hilbert space, so they commute with in this case also. The electronic Hamiltonian for a diatomic molecule is also invariant under reflections in all planes containing the internuclear line. The (X-Z) plane is such a plane, and reflection of the coordinates of the electrons in this plane corresponds to the operation $y_i \to -y_i$. If A_y is the operator that performs this reflection, then $[A_y, H] = 0$. So the Complete Set of Commuting Operators (CSCO) for a general heteronuclear diatomic molecule is $\{H, J_z, L_z, S^2, S_z, A\}$; where A is an operator that inverts only one of the two spatial co-ordinates (x or y).

In the special case of a homonuclear diatomic molecule, there is an extra symmetry since in addition to the axis of symmetry provided by the internuclear axis, there is a centre of symmetry at the midpoint of the distance between the two nuclei (the symmetry discussed in this paragraph only depends on the two nuclear charges being the same. The two nuclei can therefore have different mass, that is they can be two isotopes of the same species such as the proton and the deuteron, or O^{16} and O^{18}, and so on). Choosing this point as the origin of the coordinates, the Hamiltonian is invariant under an inversion of the coordinates of all electrons with respect to that origin, namely in the operation $\vec{r}_i \to -\vec{r}_i$. Thus the parity operator Π. Thus the CSCO for a homonuclear diatomic molecule is $\{H, J_z, L_z, S^2, S_z, A, \Pi\}$.

Molecular Term Symbol, Λ-doubling

Molecular term symbol is a shorthand expression of the group representation and angular momenta that characterize the state of a molecule. It is the equivalent of the term symbol for the atomic case. We already know the CSCO of the most general diatomic molecule. So, the good quantum numbers can sufficiently describe the state of the diatomic molecule. Here, the symmetry is explicitly stated in the nomenclature.

Angular Momentum

Here, the system is not spherically symmetric. So, $[H, L^2] \neq 0$, and the state cannot be depicted in terms of l as an eigenstate of the Hamiltonian is not an eigenstate of L^2 anymore (in contrast to the atomic term symbol, where the states were written as $^{2S+1}L_J$). But, as $[H, L_z] = 0$, the eigenvalues corresponding to L_z can still be used. If,

$$L_z|\Psi\rangle = M_L \hbar |\Psi\rangle; M_L = 0, \pm 1, \pm 2, \ldots\ldots$$
$$\Rightarrow L_z|\Psi\rangle = \pm \Lambda \hbar |\Psi\rangle; \Lambda = 0, 1, 2, \ldots\ldots$$

where $\Lambda = |M_L|$ is the absolute value (in a.u.) of the projection of the total electronic angular momentum on the internuclear axis; Λ can be used as a term symbol. By analogy with the spectroscopic notation S, P, D, F, ... used for atoms, it is customary to associate code letters with the values of Λ according to the correspondence:

$$\text{Value of } \Lambda : \quad 0 \quad 1 \quad 2 \quad 3\ldots\ldots\ldots$$
$$\updownarrow \quad \updownarrow \quad \updownarrow \quad \updownarrow$$
$$\text{Code letter:} \quad \Sigma \quad \Pi \quad \Delta \quad \Phi\ldots\ldots\ldots$$

For the individual electrons, the notation and the correspondence used are:

$\lambda = |m_l|$ and

$$\text{Value of } \lambda : \quad 0 \quad 1 \quad 2 \quad 3\ldots\ldots\ldots$$
$$\updownarrow \quad \updownarrow \quad \updownarrow \quad \updownarrow$$
$$\text{Code letter:} \quad \sigma \quad \pi \quad \delta \quad \phi\ldots\ldots\ldots$$

Axial Symmetry

Again, $[A_y, H] = 0$, and in addition: $A_y L_z = -L_z A_y$ [as $L_z = -i\hbar(x\frac{\partial}{\partial y} - y\frac{\partial}{\partial x})$]. It follows immediately that if $\Lambda \neq 0$ the action of the operator A_y on an eigenstate corresponding to the eigenvalue $\Lambda\hbar$ of L_z converts this state into another one corresponding to the eigenvalue $-\Lambda\hbar$, and that both eigenstates have the same energy. The electronic terms such that $\Lambda \neq 0$ (that is, the terms $\Pi, \Delta, \Phi, \ldots\ldots\ldots$) are thus doubly degenerate, each value of the energy corresponding to two states which differ by the direction of the projection of the orbital angular momentum along the molecular axis. This twofold degeneracy is actually only approximate and it is possible to show that the interaction between the electronic and rotational motions leads to a splitting of the terms with $\Lambda \neq 0$ into two nearby levels, which is called Λ-*doubling*.

$\Lambda = 0$ corresponds to the Σ states. These states are non-degenerate, so that the states of a Σ term can only be multiplied by a constant in a reflection through a plane containing the molecular axis. When $\Lambda = 0$, simultaneous eigenfunctions of H, L_z and A_y can be constructed. Since $A_y^2 = 1$, the eigenfunctions of A_y have eigenvalues ± 1. So to completely specify Σ states of diatomic molecules, Σ^+ states, which is left unchanged upon reflection in a plane containing the nuclei, needs to be distinguished from Σ^- states, for which it changes sign in performing that operation.

Inversion Symmetry

For a homonuclear diatomic molecule having identical nuclei both having the same charges on them (i.e., H_2, N_2, O_2............ and not O^{16} and O^{18}); there is also a centre of symmetry at the midpoint of the distance between the two nuclei. Choosing this point as the origin of the coordinates, the Hamiltonian is invariant under an inversion of the coordinates of all electrons with respect to that origin, namely in the operation $\vec{r}_i \rightarrow -\vec{r}_i$. Since the parity operator which effects this transformation also commutes with L_z, the states can be classified for given value of Λ according to their parity. The electronic states therefore split into two sets: those that are even, i.e. remain unaltered by the operation $\vec{r}_i \rightarrow -\vec{r}_i$, and those that are odd, i.e. change sign in that operation. The former are denoted by a subscript g and are called *gerade,* while the latter are denoted by a subscript u and are called *ungerade* state. The subscripts g or u are therefore added to the term symbol, so that for homonuclear diatomic molecules we have $\Sigma_g, \Sigma_u, \Pi_g, \Pi_u,$ .... So, a homonuclear diatomic molecule has four non-degenerate Σ states: $\Sigma_g^-, \Sigma_g^+, \Sigma_u^-, \Sigma_g^+$.

Spin and Total Angular Momentum

If S denotes the resultant of the individual electron spins, $s(s+1)\hbar^2$ are the eigenvalues of S and as in the case of atoms, each electronic term of the molecule is also characterised by the value of S. If spin-orbit coupling is neglected, there is a degeneracy of order $2s+1$ associated with each s for a given Λ. Just as for atoms, the quantity $2s+1$ is called the multiplicity of the term and is written as a (left) superscript, so that the term symbol is written as . For example, the symbol $^{2s+1}\Lambda$ denotes a term such that $\Lambda = 2$ and o. It is worth noting that the ground state (often labelled by the symbol X) of most diatomic molecules is such that $s = 0$ and exhibits maximum symmetry. Thus, in most cases it is a $^1\Sigma^+$ state (written as $X^1\Sigma^+$, excited states are written with $A, B, C,...$ in front) for a heteronuclear molecule and a $^1\Sigma_g^+$ state (written as $X^1\Sigma_g^+$) for a homonuclear molecule.

Spin–orbit coupling lifts the degeneracy of the electronic states. This is because the z-component of spin interacts with the z-component of the orbital angular momentum, generating a total electronic angular momentum along the molecule axis $\mathbf{J}_z$. This is characterized by the quantum number M_J, where $M_J = M_S + M_L$. Again, positive and negative values of M_J are degenerate, so the pairs (M_L, M_S) and $(-M_L,$

$-M_S$) are degenerate. These pairs are grouped together with the quantum number Ω, which is defined as the sum of the pair of values (M_L, M_S) for which M_L is positive: $\Omega = \Lambda + M_S$

Molecular Term Symbol

So, the overall molecular term symbol for the most general diatomic molecule is given by:

$$^{2S+1}\Lambda^{(+/-)}_{\Omega,(g/u)}$$

where

- S is the total spin quantum number
- Λ is the projection of the orbital angular momentum along the internuclear axis
- Ω is the projection of the total angular momentum along the internuclear axis
- $u/'g$ is the parity
- $+/-$ is the reflection symmetry along an arbitrary plane containing the internuclear axis

Von Neumann-Wigner Non-crossing Rule

Effect of Symmetry on the Matrix Elements of the Hamiltonian

The electronic terms or potential curves $E_S(R)$ of a diatomic molecule depend only on the internuclear distance R, and it is important to investigate the behaviour of these potential curves as R varies.It is of considerable interest to examine the intersection of the curves representing the different terms.

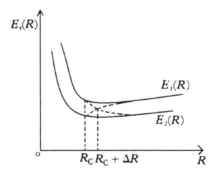

The non-crossing rule of von Neumann and Wigner. Two potential curves $E_1(R)$ and cannot $E_2(R)$ cross if the states 1 and 2 have the same symmetry

Let $E_1(R)$ and $E_2(R)$ two different electronic potential curves. If they intersect at some point, then the functions $E_1(R)$ and $E_2(R)$ will have neighbouring values near this point.

To decide whether such an intersection can occur, it is convenient to put the problem as follows. Suppose at some internuclear distance R_c the values $E_1(R_C)$ and $E_2(R_C)$ are close, but distinct (as shown in the figure). Then it is to be examined whether or not $E_1(R)$ and $E_2(R)$ can be made to intersect by the modification $R_C \rightarrow R_C + \Delta R$. The energies $E_1^{(0)} = E_1(R_C)$ and $E_2^{(0)} = E_2(R_C)$ are eigenvalues of the Hamiltonian $H_0 = H(R_C)$. The corresponding orthonormal electronic eigenstates will be denoted by $|\Phi_1^{(0)}\rangle$ and $|\Phi_2^{(0)}\rangle$ and are assumed to be real. The Hamiltonian now becomes $H \equiv H(R_C + \Delta R) = H_0 + H'$, where $H' = \frac{\partial H_0}{\partial R_C}\Delta R$ is the small perturbation operator (though it is a degenerate case, so ordinary method of perturbation won't work). setting $H'_{ij} = \langle\Phi_i^{(0)}|H'|\Phi_j^{(0)}\rangle; i, j = 1, 2$, it can be deduced that in order for $E_1(R)$ and $E_2(R)$ to be equal at the point $R_c + \Delta R$ the following two conditions are required to be fulfilled.

However, we have at our disposal only one arbitrary parameter ΔR giving the perturbation H'. Hence the

two conditions involving more than one parameter cannot in general be simultaneously satisfied (the initial assumption that $|\Phi_1^{(0)}\rangle$ and $|\Phi_2^{(0)}\rangle$ real, implies that $H'12$ is also real). So, two case can arise:

1. The matrix element $H'12$ vanishes identically. It is then possible to satisfy the first condition independently. Therefore it is possible for the crossing to occur if, for a certain value of ΔR (i.e., for a certain value of R) the first equation is satisfied. As the perturbation operator H' (or H) commutes with the symmetry operators of the molecule, this case will happen if the two states $|\Phi_1^{(0)}\rangle$ and $|\Phi_2^{(0)}\rangle$ have different symmetries (for example if they correspond to two electronic terms having different values of Λ, different parities g and u, different multiplicities, or for example are the two terms Σ^+ and Σ^-) as it can be shown that, for a scalar quantity whose operator commutes with the angular momentum and inversion operators, only the matrix elements for transitions between states of the same angular momentum and parity are non-zero and the proof remains valid, in essentially the same form, for the general case of an arbitrary symmetry operator.

2. If $|\Phi_1^{(0)}\rangle$ and $|\Phi_2^{(0)}\rangle$ have the same symmetry, then $H'12$ will in general be non-zero. Except for accidental crossing which would occur if, by coincidence, the two equations were satisfied at the same value of Λ, it is in general impossible to find a single value of Σ^+ (i.e., a single value of Σ^+) for which the two conditions are satisfied simultaneously.

Thus, in a diatomic molecule, only terms of different symmetry can intersect, while the intersection of terms of like symmetry is forbidden. This is, in general, true for any case in quantum mechanics where the Hamiltonian contains some parameter and its eigenvalues are consequently functions of that parameter. This general rule is known as von Neumann - Wigner non-crossing rule.

This general symmetry principle has important consequences is molecular spectra. In fact, in the applications of valence bond method in case of diatomic molecules, three main correspondence between the atomic and the molecular orbitals are taken care of:

1. Molecular orbitals having a given value of λ (the component of the orbital angular momentum along the internuclear axis) must connect with atomic orbitals having the same value of λ (i.e. the same value of $|m|$).

2. The parity of the wave function (g or u) must be preserved as R varies from 0 to ∞ .

3. The von Neumann-Wigner non-crossing rule must be obeyed, so that energy curves corresponding to orbitals having the same symmetry do not cross as R varies from 0 to ∞ .

Thus, von Neumann-Wigner non-crossing rule also acts as a starting point for valence bond theory.

Observable Consequences

Symmetry in diatomic molecules manifests itself directly by influencing the molecular spectra of the molecule. The effect of symmetry on different types of spectra in diatomic molecules are:

Rotational Spectrum

In the electric dipole approximation the transition amplitude for emission or absorption of radiation can be shown to be proportional to the matrix element of the electric dipole operator D. In the simplest approximation the couplings between the electronic, vibrational and rotational motions can be neglected. Disregarding spin, the complete molecular state $|\Psi_\alpha\rangle$ (corresponding to a given state α) can be broken up to a direct product of an electronic state $|\Phi_s\rangle$, vibrational state $|v\rangle$ and a rotational state $|\phi_{\Im,M_\Im,\Lambda}\rangle$ (the quantum numbers corresponding to J^2 and J_z, $\Im$ and $M_\Im$, are still good quantum numbers).The diagonal elements of D are thus given by: $D_{\alpha\alpha}=\langle\Psi_\alpha|D|\Psi_\alpha\rangle$; $\alpha\equiv(s,v,\Im,M_\Im,\Lambda)$ and are equal to the permanent electric dipole moment in the state α .

This quantity always vanishes for non-degenerate levels of atoms, because these are eigenstates of the parity operator. However, for heteronuclear diatomic molecules in which an excess of charge is associated with one of the nuclei, $D_{\alpha\alpha}$ has a finite value.

In symmetrical homonuclear diatomic molecules, the permanent electric dipole moment vanishes. Since the rotational motions (about both vertical axis and horizontal axis passing through the inversion center) preserve the symmetry of the molecule, the matrix elements of D D between different rotational states must vanish for symmetrical homonuclear diatomic molecules, unless the electronic state itself changes. As a result symmetrical molecules possess no purely rotational spectrum, without an electronic transition.

In contrast, heteronuclear diatomic molecules which possess a permanent electric dipole moment (e.g., HCl) exhibit spectra corresponding to rotational transitions, without change in the electronic state. For $\Lambda = 0$, the selection rules for a rotational transition are: $\substack{\Delta\Im = \pm 1 \\ \Delta M_\Im = 0,\pm 1}$. For $\Lambda \neq 0$, the selection rules become: $\substack{\Delta\Im = \pm 1 \\ \Delta M_\Im = 0,\pm 1}$. This is due to the fact that although the photon absorbed or emitted carries one unit of angular momentum, the nuclear rotation can change, with no change in $\Im$, if the electronic angular momentum makes an equal and opposite change. Symmetry considerations require that the electric dipole moment of a diatomic molecule is directed along the internuclear line, and this leads to the additional selection rule $\Delta\Lambda = 0$. The pure rotational spectrum of a diatomic molecule consists of lines in the far infra-red or the microwave region, the frequencies of these lines given by:

$$\hbar\omega_{\Im+1,\Im} = E_r(\Im+1) - E_r(\Im) = 2B(\Im+1)\ ;\ \text{where}\ B = \frac{\hbar^2}{2\mu R_0^2}, \text{ and } \Im \geq \Lambda$$

- Conclusion: Homonuclear diatomic molecules don't show pure rotational spectra, while heteronuclear diatomic molecules have pure rotation spectra given by the above expression.

Vibrational Spectrum

The transition matrix elements for pure vibrational transition are $\mu_{v,v'} = \langle v' | \mu | v \rangle$, where μ is the dipole moment of the diatomic molecule in the electronic state α. Because the dipole moment depends on the bond length R, its variation with displacement of the nuclei from equilibrium can be expressed as: $\mu = \mu_0 + (\frac{d\mu}{dx})_0 x + \frac{1}{2}(\frac{d^2\mu}{dx^2})_0 x^2 +$; where μ_0 is the dipole moment when the displacement is zero. The transition matrix elements are, therefore:

$$\langle v' | \mu | v \rangle = \mu_0 \langle v' | v \rangle + (\frac{d\mu}{dx})_0 \langle v' | x | v \rangle + \frac{1}{2}(\frac{d^2\mu}{dx^2})_0 \langle v' | x^2 | v \rangle + \cdots\cdots = (\frac{d\mu}{dx})_0 \langle v' | x | v \rangle + \frac{1}{2}(\frac{d^2\mu}{dx^2})_0 \langle v' | x^2 | v \rangle + \cdots\cdots$$

using orthogonality of the states. So, the transition matrix is non-zero only if the molecular dipole moment varies with displacement, for otherwise the derivatives of μ would be zero. The gross selection rule for the vibrational transitions of diatomic molecules is then: *To show a vibrational spectrum, a diatomic molecule must have a dipole moment that varies with extension. So,* homonuclear diatomic molecules do not undergo electric-dipole vibrational transitions. So, a homonuclear diatomic molecule doesn't show purely vibrational spectra.

For small displacements, the electric dipole moment of a molecule can be expected to vary linearly with the extension of the bond. This would be the case for a heteronuclear molecule in which the partial charges on the two atoms were independent of the internuclear distance. In such cases (known as harmonic approximation), the quadratic and higher terms in the expansion can be ignored and $\mu_{v,v'} = \langle v' | \mu | v \rangle = (\frac{d\mu}{dx})_0 \langle v' | x | v \rangle$. Now, the matrix elements can be expressed in position basis in terms of the harmonic oscillator wavefunctions: Hermite polynomials. Using the property of Hermite polynomials: $2(\alpha x)H_v(\alpha x) = 2vH_{v-1}(\alpha x) + H_{v+1}(\alpha x)$, it is evident that which is proportional to $|v+1\rangle$, produces two terms, one proportional to $|v+1\rangle$ and the other to $|v-1\rangle$. So, the only non-zero contributions to $\mu_{v,v'}$ comes from $v' = v \pm 1$. So, the selection rule for heteronuclear diatomic molecules is: $\Delta v = \pm 1$

- Conclusion: Homonuclear diatomic molecules show no pure vibrational spectral lines, and the vibrational spectral lines of heteronuclear diatomic molecules are governed by the above-mentioned selection rule.

Rovibrational Spectrum

Homonuclear diatomic molecules show neither pure vibrational nor pure rotational spectra. However, as the absorption of a photon requires the molecule to take up one unit of angular momentum, vibrational transitions are accompanied by a change in rotational state, which is subject to the same selection rules as for the pure rotational spectrum. For a molecule in a Σ state, the transitions between two vibration-rotation (or *rovibrational*) levels $(v, \Im)$ and $(v', \Im')$, with vibrational quantum numbers v and $v' = v+1$, fall into two sets according to whether $\Delta\Im = +1$ or $\Delta\Im = -1$. The set corresponding to $\Delta\Im = +1$ is called the *R branch*. The corresponding frequencies are given by: $\hbar\omega^R = E(v+1, \Im+1) - E(v, \Im) = 2B(\Im+1) + \hbar\omega_0;\ \Im = 0, 1, 2, \ldots\ldots$

The set corresponding to $\Delta\Im = -1$ is called the *P branch*. The corresponding frequencies are given by: $\hbar\omega^P = E(v+1, \Im-1) - E(v, \Im) = -2B\Im + \hbar\omega_0;\ \Im = 1, 2, 3, \ldots\ldots$

Both branches make up what is called a *rotational-vibrational band or a rovibrational band*. These bands are in the infra-red part of the spectrum.

If the molecule is not in a Σ state, so that $\Lambda \neq 0$, transitions with $\Delta\Im = 0$ are allowed. This gives rise to a further branch of the vibrational-rotational spectrum, called the *Q branch*. The frequencies ω^Q corresponding to the lines in this branch are given by a quadratic function of $\Im$ if B_v and B_{v+1} are unequal, and reduce to the single frequency: $\hbar\omega^Q = E(v+1, \Im) - E(v, \Im) = \hbar\omega_0$ if $B_{v+1} = B_v$.

For a heteronuclear diatomic molecule, this selection rule has two consequences:

1. Both the vibrational and rotational quantum numbers must change. The Q-branch is therefore forbidden.

2. The energy change of rotation can be either subtracted from or added to the energy change of vibration, giving the P- and R- branches of the spectrum, respectively.

Homonuclear diatomic molecules also show this kind of spectra. The selection rules, however, are a bit different.

- Conclusion: Both homo- and hetero-nuclear diatomic molecules show rovibrational spectra. A Q-branch is absent in the spectra of heteronuclear diatomic molecules.

A Special Example: Hydrogen Molecule Ion

An explicit implication of symmetry on the molecular structure can be shown in case of the simplest bi-nuclear system: a hydrogen molecule ion or a di-hydrogen cation, H_2^+. A natural trial wave function for the H_2^+ is determined by first considering the lowest-energy state of the system when the two protons are widely separated. Then there are clearly two possible states: the electron is attached either to one of the protons, forming a hydrogen atom in the ground state, or the electron is attached to the other proton, again in the ground state of a hydrogen atom (as depicted in the picture).

$$|2\rangle \qquad\qquad\qquad\qquad |1\rangle$$

Two possible initial states of the system

The trial states in the position basis (or the 'wave functions') are then:

$$\langle \mathbf{r}|1\rangle = \frac{1}{\sqrt{\pi a_0^3}} e^{-\frac{\left|\mathbf{r}-\frac{\mathbf{R}}{2}\right|}{a_0}} \text{ and } \langle \mathbf{r}|2\rangle = \frac{1}{\sqrt{\pi a_0^3}} e^{-\frac{\left|\mathbf{r}+\frac{\mathbf{R}}{2}\right|}{a_0}}$$

The analysis of H_2^+ using variational method starts assuming these forms. Again, this is only one possible combination of states. There can be other combination of states also, for example, the electron is in an excited state of the hydrogen atom. The corresponding Hamiltonian of the system is:

$$H = \frac{\mathbf{p}^2}{2m_e} - \frac{e^2}{|\mathbf{r}-\mathbf{R}/2|} - \frac{e^2}{|\mathbf{r}+\mathbf{R}/2|} + \frac{e^2}{R}$$

Clearly, using the states $|1\rangle$ and $|2\rangle$ as basis will introduce off-diagonal elements in the Hamiltonian. Here, because of the relative simplicity of the H_2^+ ion, the matrix elements can actually be calculated. Note that H_2^+ has inversion symmetry. Using its symmetry properties, we can relate the diagonal and off-diagonal elements of the Hamiltonian as:

Because $H_{11} = H_{22}$ as well as $H_{12} = H_{21}$, the linear combination of $|1\rangle$ and $|2\rangle$ that diagonalizes the Hamiltonian is $|\pm\rangle = \frac{1}{\sqrt{2\pm 2\langle 1|2\rangle}}(|1\rangle\pm|2\rangle)$ (after normalization). Now as $[H,\Pi]=0$ for H_2^+, the states $|\pm\rangle$ are also eigenstates of Π. It turns out that $|+\rangle$ and $|-\rangle$ are the eigenstates of Π with eigenvalues +1 and -1 (in other words, the wave functions $\langle \mathbf{r}|+\rangle$ and $\langle \mathbf{r}|-\rangle$ have even and odd parity, respectively). The corresponding expectation value of the energies are $E_\pm = \frac{1}{1\pm\langle 1|2\rangle}(H_{11}\pm H_{12})$.

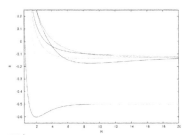

The energy vs separation graphs of H_2^+. The two lowest curves denote the E_- and E_+ sates respectively. Higher ones are the excited states. The minimum of E çorresponds to an energy -15.4 eV

From the graph, we see that only E_+ has a minimum corresponding to a separation of 1.3 Å and a total energy $E_+ = -15.4$ eV, which is less than the initial energy of the system, -13.6 eV. Thus, only the even parity state stabilizes the ion with a binding energy of 1.8 eV. As a result, the ground state of H_2^+ is $X^2\Sigma_g^+$ and this state $(|+\rangle)$ is called a bonding molecular orbital.

Thus, symmetry plays an explicit role in the formation of H_2^+.

Molecular Geometry

Molecular geometry is the three-dimensional arrangement of the atoms that constitute a molecule. It determines several properties of a substance including its reactivity, polarity, phase of matter, color, magnetism, and biological activity. The angles between bonds that an atom forms depend only weakly on the rest of molecule, i.e. they can be understood as approximately local and hence transferable properties.

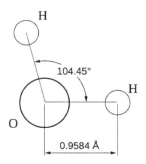

Geometry of the water molecule

Determination

The molecular geometry can be determined by various spectroscopic methods and diffraction methods. IR, microwave and Raman spectroscopy can give information about the molecule geometry from the details of the vibrational and rotational absorbance detected by these techniques. X-ray crystallography, neutron diffraction and electron diffraction can give molecular structure for crystalline solids based on the distance between nuclei and concentration of electron density. Gas electron diffraction can be used for small molecules in the gas phase. NMR and FRET methods can be used to determine complementary information including relative distances, dihedral angles, angles, and connectivity. Molecular geometries are best determined at low temperature because at higher temperatures the molecular structure is averaged over more accessible geometries. Larger molecules often exist in multiple stable geometries (conformational isomerism) that are close in energy on the potential energy surface. Geometries can also be computed by ab initio quantum chemistry methods to high accuracy. The molecular geometry can be different as a solid, in solution, and as a gas.

The position of each atom is determined by the nature of the chemical bonds by which it is connected to its neighboring atoms. The molecular geometry can be described by the positions of these atoms in space, evoking bond lengths of two joined atoms, bond angles of three connected atoms, and torsion angles (dihedral angles) of three consecutive bonds.

The Influence of Thermal Excitation

Since the motions of the atoms in a molecule are determined by quantum mechanics, one must define "motion" in a quantum mechanical way. The overall (external) quantum mechanical motions translation and rotation hardly change the geometry of the molecule. (To some extent rotation influences the geometry via Coriolis forces and centrifugal distortion, but this is negligible for the present discussion.) In addition to translation and rotation, a third type of motion is molecular vibration, which corresponds to internal motions of the atoms such as bond stretching and bond angle variation. The molecular vibrations are harmonic (at least to good approximation), and the atoms oscillate about their equilibrium positions, even at the absolute zero of temperature. At absolute zero all atoms are in their vibrational ground state and show zero point quantum mechanical motion, so that the wavefunction of a single vibrational mode is not a sharp peak, but an exponential of finite width (the wavefunction for $n = 0$ depicted in the article on the quantum harmonic oscillator). At higher temperatures the vibrational modes may be thermally excited (in a classical interpretation one expresses this by stating that "the molecules will vibrate faster"), but they oscillate still around the recognizable geometry of the molecule.

To get a feeling for the probability that the vibration of molecule may be thermally excited, we inspect the Boltzmann factor $\beta = \exp\left(-\dfrac{\Delta E}{kT}\right)$, where ΔE is the excitation energy of the vibrational mode, k the Boltzmann constant and T the absolute temperature. At 298 K (25 °C), typical values for the Boltzmann factor β are: $\beta = 0.089$ for $\Delta E = 500$ cm^{-1}; $\beta = 0.008$ for $\Delta E = 1000$ cm^{-1}; $\beta = 7\times10^{-4}$ for $\Delta E = 1500$ cm^{-1}. (The reciprocal centimeter is an energy unit that is commonly used in infrared spectroscopy; 1 cm^{-1} corresponds to 1.23984×10^{-4} eV). When an excitation energy is 500 cm^{-1}, then about 8.9 percent of the molecules are thermally excited at room temperature. To put this in perspective: the lowest excitation vibrational energy in water is the bending mode (about 1600 cm^{-1}). Thus, at room temperature less than 0.07 percent of all the molecules of a given amount of water will vibrate faster than at absolute zero.

As stated above, rotation hardly influences the molecular geometry. But, as a quantum mechanical motion, it is thermally excited at relatively (as compared to vibration) low temperatures. From a classical point of view it can be stated that at higher temperatures more molecules will rotate faster, which implies that they have higher angular velocity and angular momentum. In quantum mechanical language: more eigenstates of higher angular momentum become thermally populated with rising temperatures. Typical rotational excitation energies are on the order of a few cm^{-1}. The results of many spectroscopic experiments are broadened because they involve an averaging over rotational states. It is often difficult to extract geometries from spectra at high temperatures, because the number of rotational states probed in the experimental averaging increases with increasing temperature. Thus, many spectroscopic observations can only be expected to yield reliable molecular geometries at temperatures close to absolute zero, because at higher temperatures too many higher rotational states are thermally populated.

Bonding

Molecules, by definition, are most often held together with covalent bonds involving single, double, and/or triple bonds, where a "bond" is a shared pair of electrons (the other method of bonding between atoms is called ionic bonding and involves a positive cation and a negative anion).

Molecular geometries can be specified in terms of bond lengths, bond angles and torsional angles. The bond length is defined to be the average distance between the nuclei of two atoms bonded together in any given molecule. A bond angle is the angle formed between three atoms across at least two bonds. For four atoms bonded together in a chain, the torsional angle is the angle between the plane formed by the first three atoms and the plane formed by the last three atoms.

There exists a mathematical relationship among the bond angles for one central atom and four peripheral atoms (labeled 1 through 4) expressed by the following determinant. This constraint removes one degree of freedom from the choices of (originally) six free bond angles to leave only five choices of bond angles. (Note that the angles θ_{11}, θ_{22}, θ_{33}, and θ_{44} are always zero and that this relationship can be modified for a different number of peripheral atoms by expanding/contracting the square matrix.)

$$0 = \begin{vmatrix} \cos\theta_{11} & \cos\theta_{12} & \cos\theta_{13} & \cos\theta_{14} \\ \cos\theta_{21} & \cos\theta_{22} & \cos\theta_{23} & \cos\theta_{24} \\ \cos\theta_{31} & \cos\theta_{32} & \cos\theta_{33} & \cos\theta_{34} \\ \cos\theta_{41} & \cos\theta_{42} & \cos\theta_{43} & \cos\theta_{44} \end{vmatrix}$$

Molecular geometry is determined by the quantum mechanical behavior of the electrons. Using the valence bond approximation this can be understood by the type of bonds between the atoms that make up the molecule. When atoms interact to form a chemical bond, the atomic orbitals of each atom are said to combine in a process called orbital hybridisation. The two most common types of bonds are sigma bonds (usually formed by hybrid orbitals) and pi bonds (formed by unhybridized p orbitals for atoms of main group elements). The geometry can also be understood by molecular orbital theory where the electrons are delocalised.

An understanding of the wavelike behavior of electrons in atoms and molecules is the subject of quantum chemistry.

Isomers

Isomers are types of molecules that share a chemical formula but have different geometries, resulting in very different properties:

- A pure substance is composed of only one type of isomer of a molecule (all have the same geometrical structure).

- Structural isomers have the same chemical formula but different physical arrangements, often forming alternate molecular geometries with very different properties. The atoms are not bonded (connected) together in the same orders.

  o Functional isomers are special kinds of structural isomers, where certain groups of atoms exhibit a special kind of behavior, such as an ether or an alcohol.

- Stereoisomers may have many similar physicochemical properties (melting point, boiling point) and at the same time very different biochemical activities. This is because they exhibit a handedness that is commonly found in living systems. One manifestation of this chirality or handedness is that they have the ability to rotate polarized light in different directions.

- Protein folding concerns the complex geometries and different isomers that proteins can take.

Types of Molecular Structure

A bond angle is the geometric angle between two adjacent bonds. Some common shapes of simple molecules include:

- Linear: In a linear model, atoms are connected in a straight line. The bond angles are set at 180°. For example, carbon dioxide and nitric oxide have a linear molecular shape.

- Trigonal planar: Molecules with the trigonal planar shape are somewhat triangular and in one plane (flat). Consequently, the bond angles are set at 120°. For example, boron trifluoride.

- Bent: Bent or angular molecules have a non-linear shape. For example, water (H_2O), which has an angle of about 105°. A water molecule has two pairs of bonded electrons and two unshared lone pairs.

- Tetrahedral: *Tetra-* signifies four, and *-hedral* relates to a face of a solid, so "tetrahedral" literally means "having four faces". This shape is found when there are four bonds all on one central atom, with no extra unshared electron pairs. In accordance with the VSEPR (valence-shell electron pair repulsion theory), the bond angles between the electron bonds are arccos(−1/3) = 109.47°. For example, methane (CH_4) is a tetrahedral molecule.

- Octahedral: *Octa-* signifies eight, and *-hedral* relates to a face of a solid, so "octahedral" means "having eight faces". The bond angle is 90 degrees. For example, sulfur hexafluoride (SF_6) is an octahedral molecule.

- Trigonal pyramidal: A trigonal pyramidal molecule has a pyramid-like shape with a triangular base. Unlike the linear and trigonal planar shapes but similar to the tetrahedral orientation, pyramidal shapes require three dimensions in order to fully separate the electrons. Here, there are only three pairs of bonded electrons, leaving one unshared lone pair. Lone pair – bond pair repulsions change the bond angle from the tetrahedral angle to a slightly lower value. For example, ammonia (NH_3).

VSEPR Table

The bond angles in the table below are ideal angles from the simple VSEPR theory, followed by the actual angle for the example given in the following column where this differs. For many cases, such as trigonal pyramidal and bent, the actual angle for the example differs from the ideal angle, but all examples differ by different amounts. For example, the angle in H_2S (92°) differs from the tetrahedral angle by much more than the angle for H_2O (104.48°) does.

Bonding electron pairs	Lone pairs	Electron domains (Steric number)	Shape	Ideal bond angle (example's bond angle)	Example	Image
2	0	2	linear	180°	CO_2	
3	0	3	trigonal planar	120°	BF_3	

2	1	3	bent	120° (119°)	SO_2	
4	0	4	tetrahedral	109.5°	CH_4	
3	1	4	trigonal pyramidal	109.5 (107.8°)	NH_3	
2	2	4	bent	109.5° (104.48°)	H_2O	
5	0	5	trigonal bipyramidal	90°, 120°, 180°	PCl_5	
4	1	5	seesaw	ax–ax 180° (173.1°), eq–eq 120° (101.6°), ax–eq 90°	SF_4	
3	2	5	T-shaped	90° (87.5°), 180° (175°)	ClF_3	
2	3	5	linear	180°	XeF_2	
6	0	6	octahedral	90°, 180°	SF_6	
5	1	6	square pyramidal	90° (84.8°)	BrF_5	
4	2	6	square planar	90°, 180°	XeF_4	
7	0	7	pentagonal bipyramidal	90°, 72°, 180°	IF_7	
6	1	7	pentagonal pyramidal	72°, 90°, 144°	$XeOF_5^-$	
5	2	7	planar pentagonal	72°, 144°	XeF_5^-	
8	0	8	square antiprismatic		XeF_8^{2-}	

9	0	9	tricapped trigonal prismatic		ReH_9^{2-}	

3D Representations

- Line or stick – atomic nuclei are not represented, just the bonds as sticks or lines. As in 2D molecular structures of this type, atoms are implied at each vertex.

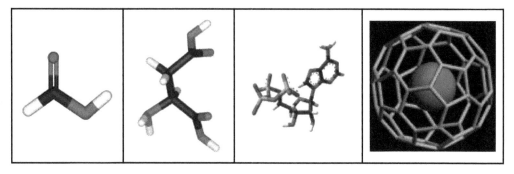

- Electron density plot – shows the electron density determined either crystallographically or using quantum mechanics rather than distinct atoms or bonds.

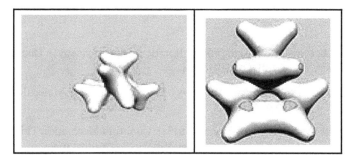

- Ball and stick – atomic nuclei are represented by spheres (balls) and the bonds as sticks.

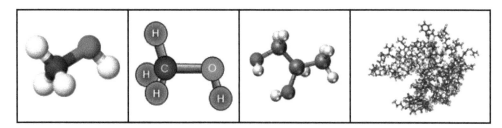

- Spacefilling models or CPK models (also an atomic coloring scheme in representations) – the molecule is represented by overlapping spheres representing the atoms.

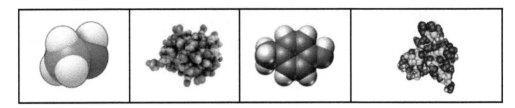

- Cartoon – a representation used for proteins where loops, beta sheets, alpha helices are represented diagrammatically and no atoms or bonds are represented explicitly just the protein backbone as a smooth pipe.

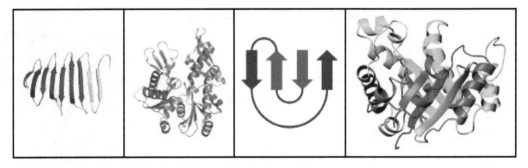

The greater the amount of lone pairs contained in a molecule the smaller the angles between the atoms of that molecule. The VSEPR theory predicts that lone pairs repel each other, thus pushing the different atoms away from them.

References

- Brown, John M.; Carrington, Alan (2003). Rotational Spectroscopy of Diatomic Molecules. Cambridge University Press. ISBN 0521530784.

- B.H. Bransden, ,C.J. Joachain (24 Apr 2003). Physics of Atoms & Molecules (2nd edition). Prentice Hall. ISBN 978-8177582796.

- L. D. Landau, & L. M. Lifshitz (January 1, 1981). Quantum Mechanics, Third Edition: Non-Relativistic Theory (Volume 3). Pergamon Press. ISBN 978-0750635394.

- Townsend, John S. A Modern Approach to Quantum Mechanics (2nd edition). University Science Books. ISBN 978-1891389788.

- Cotton, F. Albert; Wilkinson, Geoffrey; Murillo, Carlos A.; Bochmann, Manfred (1999), Advanced Inorganic Chemistry (6th ed.), New York: Wiley-Interscience, ISBN 0-471-19957-5

Density Functional Theory

To determine the solution of the Schrödinger equation for an N-body system, the density functional theory is employed. Electron density is used as the functional or the underlying property and the electron density functional in turn helps determine the total energy of the system. This chapter explores the density functional theory, its contributing factors like Lieb–Oxford inequality, Local-density approximation and the Minnesota functionals; the Runge–Gross theorem which forms a basis for the time-dependent density functional theory and the orbital-free density functional theory.

Electronic Density

In quantum mechanics, and in particular quantum chemistry, the electronic density is a measure of the probability of an electron occupying an infinitesimal element of space surrounding any given point. It is a scalar quantity depending upon three spatial variables and is typically denoted as either $\rho(\mathbf{r})$ or $n(\mathbf{r})$. The density is determined, through definition, by the normalized N-electron wavefunction which itself depends upon $4N$ variables ($3N$ spatial and N spin coordinates). Conversely, the density determines the wave function modulo a phase factor, providing the formal foundation of density functional theory.

Definition

The electronic density corresponding to a normalized N-electron wavefunction (with $\mathbf{r}$ and s denoting spatial and spin variables respectively) is defined as

$$\rho(\mathbf{r}) = N \sum_{s_1} \cdots \sum_{s_N} \int d\mathbf{r}_2 \cdots \int d\mathbf{r}_N \, | \Psi(\mathbf{r}, s_1, \mathbf{r}_2, s_2, ..., \mathbf{r}_N, s_N) |^2,$$

$$= \langle \Psi | \hat{\rho}(\mathbf{r}) | \Psi \rangle,$$

where the operator corresponding to the density observable is

$$\hat{\rho}(\mathbf{r}) = \sum_{i=1}^{N} \sum_{s_i} \delta(\mathbf{r} - \mathbf{r}_i).$$

In Hartree–Fock and density functional theories the wave function is typically represented as a single Slater determinant constructed from N orbitals, φ_k, with corresponding occupations n_k. In these situations the density simplifies to

$$\rho(\mathbf{r}) = \sum_{k=1}^{N} n_k \, |\varphi_k(\mathbf{r})|^2 \, .$$

General Properties

From its definition, the electron density is a non-negative function integrating to the total number of electrons. Further, for a system with kinetic energy T, the density satisfies the inequalities

$$\frac{1}{2} \int d\mathbf{r} \, \left(\nabla \sqrt{\rho(\mathbf{r})}\right)^2 \leq T.$$

$$\frac{3}{2} \left(\frac{\pi}{2}\right)^{4/3} \left(\int d\mathbf{r} \, \rho^3(\mathbf{r})\right)^{1/3} \leq T.$$

For finite kinetic energies, the first (stronger) inequality places the square root of the density in the Sobolev space $H^1(\mathbf{R}^3)$. Together with the normalization and non-negativity this defines a space containing physically acceptable densities as

$$\mathcal{J}_N = \left\{\rho \,\middle|\, \rho(\mathbf{r}) \geq 0, \, \rho^{1/2}(\mathbf{r}) \in H^1(\mathbf{R}^3), \int d\mathbf{r} \, \rho(\mathbf{r}) = N\right\}.$$

The second inequality places the density in the L^3 space. Together with the normalization property places acceptable densities within the intersection of L^1 and L^3 – a superset of $\mathcal{J}_N$.

Topology

The ground state electronic density of an atom is conjectured to be a monotonically decaying function of the distance from the nucleus.

Nuclear Cusp Condition

The electronic density displays cusps at each nucleus in a molecule as a result of the unbounded electron-nucleus Coulomb potential. This behavior is quantified by the Kato cusp condition formulated in terms of the spherically averaged density, $\bar{\rho}$, about any given nucleus as

$$\left.\frac{\partial}{\partial r_\alpha} \bar{\rho}(r_\alpha)\right|_{r_\alpha = 0} = -2Z_\alpha \bar{\rho}(0).$$

That is, the radial derivative of the spherically averaged density, evaluated at any nucleus, is equal to twice the density at that nucleus multiplied by the negative of the atomic number (Z).

Asymptotic Behavior

The nuclear cusp condition provides the near-nuclear (small r) density behavior as

$$\rho(r) \sim e^{-2Z_a r}.$$

The long-range (large r) behavior of the density is also known, taking the form

$$\rho(r) \sim e^{-2\sqrt{2I} r}.$$

where I is the ionization energy of the system.

Response Density

Another more-general definition of a density is the "linear-response density". This is the density that when contracted with any spin-free, one-electron operator yields the associated property defined as the derivative of the energy. For example, a dipole moment is the derivative of the energy with respect to an external magnetic field and is not the expectation value of the operator over the wavefunction. For some theories they are the same when the wavefunction is converged. The occupation numbers are not limited to the range of zero to two, and therefore sometimes even the response density can be negative in certain regions of space.

Density Functional Theory

Density functional theory (DFT) is a computational quantum mechanical modelling method used in physics, chemistry and materials science to investigate the electronic structure (principally the ground state) of many-body systems, in particular atoms, molecules, and the condensed phases. Using this theory, the properties of a many-electron system can be determined by using functionals, i.e. functions of another function, which in this case is the spatially dependent electron density. Hence the name density functional theory comes from the use of functionals of the electron density. DFT is among the most popular and versatile methods available in condensed-matter physics, computational physics, and computational chemistry.

DFT has been very popular for calculations in solid-state physics since the 1970s. However, DFT was not considered accurate enough for calculations in quantum chemistry until the 1990s, when the approximations used in the theory were greatly refined to better model the exchange and correlation interactions. Computational costs are relatively low when compared to traditional methods, such as Hartree–Fock theory and its descendants based on the complex many-electron wavefunction.

Despite recent improvements, there are still difficulties in using density functional theory to properly describe intermolecular interactions (of critical importance to understanding chemical reactions), especially van der Waals forces (dispersion); charge transfer excitations; transition states, global potential energy surfaces, dopant interactions and some other strongly correlated systems; and in calculations of the band gap and ferromagnetism in semiconductors. Its incomplete treatment of dispersion can adversely affect the accuracy of DFT (at least when used alone and uncorrected) in the treatment of systems which are dominated by dispersion (e.g. interacting noble gas atoms) or where dispersion competes significantly with other effects (e.g. in biomolecules). The development of new DFT methods designed to overcome this problem, by alterations to the functional and inclusion of additional terms to account for both core and valence electrons or by the inclusion of additive terms, is a current research topic.

Overview of Method

Although density functional theory has its conceptual roots in the Thomas–Fermi model, DFT was put on a firm theoretical footing by the two Hohenberg–Kohn theorems (H–K). The original H–K theorems held only for non-degenerate ground states in the absence of a magnetic field, although they have since been generalized to encompass these.

The first H–K theorem demonstrates that the ground state properties of a many-electron system are uniquely determined by an electron density that depends on only 3 spatial coordinates. It lays the groundwork for reducing the many-body problem of N electrons with 3N spatial coordinates to 3 spatial coordinates, through the use of functionals of the electron density. This theorem can be extended to the time-dependent domain to develop time-dependent density functional theory (TDDFT), which can be used to describe excited states.

The second H–K theorem defines an energy functional for the system and proves that the correct ground state electron density minimizes this energy functional.

Within the framework of Kohn–Sham DFT (KS DFT), the intractable many-body problem of interacting electrons in a static external potential is reduced to a tractable problem of non-interacting electrons moving in an effective potential. The effective potential includes the external potential and the effects of the Coulomb interactions between the electrons, e.g., the exchange and correlation interactions. Modeling the latter two interactions becomes the difficulty within KS DFT. The simplest approximation is the local-density approximation (LDA), which is based upon exact exchange energy for a uniform electron gas, which can be obtained from the Thomas–Fermi model, and from fits to the correlation energy for a uniform electron gas. Non-interacting systems are relatively easy to solve as the wavefunction can be represented as a Slater determinant of orbitals. Further, the kinetic energy functional of such a system is known exactly. The exchange-correlation part of the total-energy functional remains unknown and must be approximated.

Another approach, less popular than KS DFT but arguably more closely related to the spirit of the original H-K theorems, is orbital-free density functional theory (OFDFT), in which approximate functionals are also used for the kinetic energy of the non-interacting system.

Derivation and Formalism

As usual in many-body electronic structure calculations, the nuclei of the treated molecules or clusters are seen as fixed (the Born–Oppenheimer approximation), generating a static external potential V in which the electrons are moving. A stationary electronic state is then described by a wavefunction $\Psi(\vec{r}_1,\ldots,\vec{r}_N)$ satisfying the many-electron time-independent Schrödinger equation

$$\hat{H}\Psi = \left[\hat{T} + \hat{V} + \hat{U}\right]\Psi = \left[\sum_i^N \left(-\frac{\hbar^2}{2m_i}\nabla_i^2\right) + \sum_i^N V(\vec{r}_i) + \sum_{i<j}^N U(\vec{r}_i,\vec{r}_j)\right]\Psi = E\Psi$$

where, for the $N--$-electron system, $\hat{H}$ is the Hamiltonian, E is the total energy, $\hat{T}$ is the kinetic energy, $\hat{V}$ is the potential energy from the external field due to positively charged nuclei, and $\hat{U}$ is the electron-electron interaction energy. The operators $\hat{T}$ and $\hat{U}$ are called universal operators as they are the same for any N-electron system, while $\hat{V}$ is system dependent. This complicated many-particle equation is not separable into simpler single-particle equations because of the interaction term $\hat{U}..$

There are many sophisticated methods for solving the many-body Schrödinger equation based on the expansion of the wavefunction in Slater determinants. While the simplest one is the Hartree–Fock method, more sophisticated approaches are usually categorized as post-Hartree–Fock methods. However, the problem with these methods is the huge computational effort, which makes it virtually impossible to apply them efficiently to larger, more complex systems.

Here DFT provides an appealing alternative, being much more versatile as it provides a way to systematically map the many-body problem, with $\hat{U}$, onto a single-body problem without $\hat{U}$. . In DFT the key variable is the electron density $n(\vec{r})$, which for a normalized ψ is given by

$$n(\vec{r}) = N\int d^3r_2 \cdots \int d^3r_N \Psi^*(\vec{r},\vec{r}_2,\ldots,\vec{r}_N)\Psi(\vec{r},\vec{r}_2,\ldots,\vec{r}_N).$$

This relation can be reversed, i.e., for a given ground-state density $n_0(\vec{r})$ it is possible, in principle, to calculate the corresponding ground-state wavefunction $\emptyset_0(\vec{r}_1,\ldots,\vec{r}_N)$. In other words, ψ is a unique functional of n_0,

$$\Psi_0 = \Psi[n_0]$$

and consequently the ground-state expectation value of an observable $\hat{O}$ is also a functional of n_0

$$O[n_0] = \left\langle \Psi[n_0]\middle|\hat{O}\middle|\Psi[n_0]\right\rangle.$$

In particular, the ground-state energy is a functional of n_0

$$E_0 = E[n_0] = \left\langle \Psi[n_0]\middle|\hat{T}+\hat{V}+\hat{U}\middle|\Psi[n_0]\right\rangle$$

where the contribution of the external potential $\langle \Psi[n_0] | \hat{V} | \Psi[n_0] \rangle$ can be written explicitly in terms of the ground-state density n_0

$$V[n_0] = \int V(\vec{r}) n_0(\vec{r}) \mathrm{d}^3 r$$

More generally, the contribution of the external potential $\langle \Psi | \hat{V} | \Psi \rangle$ can be written explicitly in terms of the density n_0

$$V[n] = \int V(\vec{r}) n(\vec{r}) \mathrm{d}^3 r$$

The functionals T[n] and U[n] are called universal functionals, while V[n] is called a non-universal functional, as it depends on the system under study. Having specified a system, i.e., having specified $\hat{V}$, one then has to minimize the functional

$$E[n] = T[n] + U[n] + \int V(\vec{r}) n(\vec{r}) \mathrm{d}^3 r$$

with respect to $n(\vec{r})$, assuming one has got reliable expressions for $T[n]$ and $U[n]$. A successful minimization of the energy functional will yield the ground-state density n_0 and thus all other ground-state observables

The variational problems of minimizing the energy functional $E[n]$ can be solved by applying the Lagrangian method of undetermined multipliers. First, one considers an energy functional that doesn't explicitly have an electron-electron interaction energy term,

$$E_s[n] = \left\langle \Psi_s[n] \left| \hat{T} + \hat{V}_s \right| \Psi_s[n] \right\rangle$$

where $\hat{T}$ denotes the kinetic energy operator and $\hat{V}_s$ is an external effective potential in which the particles are moving, so that $n_s(\vec{r}) \overset{\text{def}}{=} n(\vec{r})$.

Thus, one can solve the so-called Kohn–Sham equations of this auxiliary non-interacting system,

$$\left[-\frac{\hbar^2}{2m} \nabla^2 + V_s(\vec{r}) \right] \phi_i(\vec{r}) = \epsilon_i \phi_i(\vec{r})$$

which yields the orbitals Φi that reproduce the density $n(\vec{r})$ of the original many-body system

$$n(\vec{r}) \overset{\text{def}}{=} n_s(\vec{r}) = \sum_i^N \left| \phi_i(\vec{r}) \right|^2$$

The effective single-particle potential can be written in more detail as

$$V_s(\vec{r}) = V(\vec{r}) + \int \frac{e^2 n_s(\vec{r}')}{|\vec{r} - \vec{r}'|} \mathrm{d}^3 r' + V_{\mathrm{XC}}[n_s(\vec{r})]$$

where the second term denotes the so-called Hartree term describing the electron-electron Coulomb repulsion, while the last term V_{XC} is called the exchange-correlation potential. Here, V_{XC} includes all the many-particle interactions. Since the Hartree term and V_{xc} depend on $n(\vec{r})$, which depends on the Φi, which in turn depend on Vs, the problem of solving the Kohn–Sham equation has to be done in a self-consistent (i.e., iterative) way. Usually one starts with an initial guess for $n(\vec{r})$, then calculates the corresponding Vs and solves the Kohn–Sham equations for the Φi. From these one calculates a new density and starts again. This procedure is then repeated until convergence is reached. A non-iterative approximate formulation called Harris functional DFT is an alternative approach to this.

Relativistic Density Functional Theory (Explicit Functional Forms)

The same theorems can be proven in the case of relativistic electrons thereby providing generalization of DFT for relativistic case. Unlike nonrelativistic theory in relativistic case it's possible to derive a few exact and explicit formulas for relativistic density functional.

Let's consider an electron in hydrogen-like ion obeying relativistic Dirac equation. Hamiltonian H for relativistic electron moving in the Coulomb potential can be chosen in the following form (atomic units are used):

$$H = c(\vec{\alpha} \cdot \vec{p}) + eV + mc^2 \beta,$$

where $V = -\frac{eZ}{r}$ is Coulomb potential of a point-like nucleus, $\vec{p}$ is a momentum operator of electron, e, m and c are electron electric charge, mass and speed of light in vacuum constants respectively.

$$\vec{\alpha} = \begin{pmatrix} 0 & \vec{\sigma} \\ \vec{\sigma} & 0 \end{pmatrix}, \beta = \begin{pmatrix} I & 0 \\ 0 & -I \end{pmatrix}.$$

To find out eigen functions and corresponding energies one solves the eigen function equation:

$$.H\Psi = E\Psi,$$

where $\Psi = (\Psi(1), \Psi(2), \Psi(3), \Psi(4))^T$ is a four component wave function and E is associated eigen energy. It is demonstrated in the article that application of the virial theorem to eigen function equation produces the following formula for eigen energy of any bound state.

and analogously the virial theorem applied to the eigen function equation with squared Hamiltonian:

$$E = mc^2 \langle \Psi | \beta | \Psi \rangle = mc^2 \int |\Psi(1)|^2 + |\Psi(2)|^2 - |\Psi(3)|^2 - |\Psi(4)|^2 \, d\tau,$$

It's easy to see that both written above formulas represent density functionals. The former formula can be easily generalized for multi-electron case.

Approximations (Exchange-correlation Functionals)

The major problem with DFT is that the exact functionals for exchange and correlation are not known except for the free electron gas. However, approximations exist which permit the calculation of certain physical quantities quite accurately. In physics the most widely used approximation is the local-density approximation (LDA), where the functional depends only on the density at the coordinate where the functional is evaluated:

$$E_{XC}^{LDA}[n] = \int \epsilon_{XC}(n) n(\vec{r}) d^3 r.$$

The local spin-density approximation (LSDA) is a straightforward generalization of the LDA to include electron spin:

$$E_{XC}^{LSDA}[n_\uparrow, n_\downarrow] = \int \epsilon_{XC}(n_\uparrow, n_\downarrow) n(\vec{r}) d^3 r.$$

Highly accurate formulae for the exchange-correlation energy density $\partial_{XC}(n_\uparrow, n_\downarrow)$ have been constructed from quantum Monte Carlo simulations of jellium.

The LDA assumes that the density is the same everywhere. Because of this, the LDA has a tendency to over-estimate the exchange-correlation energy. To correct for this tendency, it is common to expand in terms of the gradient of the density in order to account for the non-homogeneity of the true electron density. This allows for corrections based on the changes in density away from the coordinate. These expansions are referred to as generalized gradient approximations (GGA) and have the following form:

$$E_{XC}^{GGA}[n_\uparrow, n_\downarrow] = \int \epsilon_{XC}(n_\uparrow, n_\downarrow, \vec{\nabla} n_\uparrow, \vec{\nabla} n_\downarrow) n(\vec{r}) d^3 r.$$

Using the latter (GGA), very good results for molecular geometries and ground-state energies have been achieved.

Potentially more accurate than the GGA functionals are the meta-GGA functionals, a natural development after the GGA (generalized gradient approximation). Meta-GGA DFT functional in its original form includes the second derivative of the electron density (the Laplacian) whereas GGA includes only the density and its first derivative in the exchange-correlation potential.

Functionals of this type are, for example, TPSS and the Minnesota Functionals. These functionals include a further term in the expansion, depending on the density, the gradient of the density and the Laplacian (second derivative) of the density.

Difficulties in expressing the exchange part of the energy can be relieved by including a component of the exact exchange energy calculated from Hartree–Fock theory. Functionals of this type are known as hybrid functionals.

Generalizations to Include Magnetic Fields

The DFT formalism described above breaks down, to various degrees, in the presence of a vector potential, i.e. a magnetic field. In such a situation, the one-to-one mapping between the ground-state electron density and wavefunction is lost. Generalizations to include the effects of magnetic fields have led to two different theories: current density functional theory (CDFT) and magnetic field density functional theory (BDFT). In both these theories, the functional used for the exchange and correlation must be generalized to include more than just the electron density. In current density functional theory, developed by Vignale and Rasolt, the functionals become dependent on both the electron density and the paramagnetic current density. In magnetic field density functional theory, developed by Salsbury, Grayce and Harris, the functionals depend on the electron density and the magnetic field, and the functional form can depend on the form of the magnetic field. In both of these theories it has been difficult to develop functionals beyond their equivalent to LDA, which are also readily implementable computationally. Recently an extension by Pan and Sahni extended the Hohenberg-Kohn theorem for non constant magnetic fields using the density and the current density as fundamental variables.

Applications

In general, density functional theory finds increasingly broad application in the chemical and material sciences for the interpretation and prediction of complex system behavior at an atomic scale. Specifically, DFT computational methods are applied for the study of systems to synthesis and processing parameters. In such systems, experimental studies are often encumbered by inconsistent results and non-equilibrium conditions. Examples of contemporary DFT applications include studying the effects of dopants on phase transformation behavior in oxides, magnetic behaviour in dilute magnetic semiconductor materials and the study of magnetic and electronic behavior in ferroelectrics and dilute magnetic semiconductors. Also, it has been shown that DFT has a good results in the prediction of sensitivity of some nanostructures to environment pollutants like SO_2 or Acrolein as well as prediction of mechanical properties.

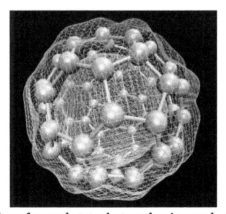

C_{60} with isosurface of ground-state electron density as calculated with DFT.

In practice, Kohn–Sham theory can be applied in several distinct ways depending on what is being investigated. In solid state calculations, the local density approximations are still commonly used along with plane wave basis sets, as an electron gas approach is more appropriate for electrons delocalised through an infinite solid. In molecular calculations, however, more sophisticated functionals are needed, and a huge variety of exchange-correlation functionals have been developed for chemical applications. Some of these are inconsistent with the uniform electron gas approximation, however, they must reduce to LDA in the electron gas limit. Among physicists, probably the most widely used functional is the revised Perdew–Burke–Ernzerhof exchange model (a direct generalized-gradient parametrization of the free electron gas with no free parameters); however, this is not sufficiently calorimetrically accurate for gas-phase molecular calculations. In the chemistry community, one popular functional is known as BLYP (from the name Becke for the exchange part and Lee, Yang and Parr for the correlation part). Even more widely used is B3LYP which is a hybrid functional in which the exchange energy, in this case from Becke's exchange functional, is combined with the exact energy from Hartree–Fock theory. Along with the component exchange and correlation functionals, three parameters define the hybrid functional, specifying how much of the exact exchange is mixed in. The adjustable parameters in hybrid functionals are generally fitted to a 'training set' of molecules. Unfortunately, although the results obtained with these functionals are usually sufficiently accurate for most applications, there is no systematic way of improving them (in contrast to some of the traditional wavefunction-based methods like configuration interaction or coupled cluster theory). Hence in the current DFT approach it is not possible to estimate the error of the calculations without comparing them to other methods or experiments.

Thomas–Fermi Model

The predecessor to density functional theory was the Thomas–Fermi model, developed independently by both Thomas and Fermi in 1927. They used a statistical model to approximate the distribution of electrons in an atom. The mathematical basis postulated that electrons are distributed uniformly in phase space with two electrons in every h^3 of volume. For each element of coordinate space volume $d^3 r$ we can fill out a sphere of momentum space up to the Fermi momentum p_f

$$\frac{4}{3}\pi p_f^3(\vec{r}).$$

Equating the number of electrons in coordinate space to that in phase space gives:

$$n(\vec{r}) = \frac{8\pi}{3h^3} p_f^3(\vec{r}).$$

Solving for p_f and substituting into the classical kinetic energy formula then leads directly to a kinetic energy represented as a functional of the electron density:

$$t_{TF}[n] = \frac{p^2}{2m_e} \propto \frac{(n^{\frac{1}{3}})^2}{2m_e} \propto n^{\frac{2}{3}}(\vec{r})$$

$$T_{TF}[n] = C_F \int n(\vec{r}) n^{\frac{2}{3}}(\vec{r}) d^3r = C_F \int n^{\frac{5}{3}}(\vec{r}) d^3r$$

where $C_F = \frac{3h^2}{10m_e}(\frac{3}{8\pi})^{\frac{2}{3}}$.

As such, they were able to calculate the energy of an atom using this kinetic energy functional combined with the classical expressions for the nuclear-electron and electron-electron interactions (which can both also be represented in terms of the electron density).

Although this was an important first step, the Thomas–Fermi equation's accuracy is limited because the resulting kinetic energy functional is only approximate, and because the method does not attempt to represent the exchange energy of an atom as a conclusion of the Pauli principle. An exchange energy functional was added by Dirac in 1928.

However, the Thomas–Fermi–Dirac theory remained rather inaccurate for most applications. The largest source of error was in the representation of the kinetic energy, followed by the errors in the exchange energy, and due to the complete neglect of electron correlation.

Teller (1962) showed that Thomas–Fermi theory cannot describe molecular bonding. This can be overcome by improving the kinetic energy functional.

The kinetic energy functional can be improved by adding the Weizsäcker (1935) correction:

$$T_W[n] = \frac{\hbar^2}{8m} \int \frac{|\nabla n(\vec{r})|^2}{n(\vec{r})} d^3r.$$

Pseudo-potentials

The many electron Schrödinger equation can be very much simplified if electrons are divided in two groups: valence electrons and inner core electrons. The electrons in the inner shells are strongly bound and do not play a significant role in the chemical binding of atoms; they also partially screen the nucleus, thus forming with the nucleus an almost inert core. Binding properties are almost completely due to the valence electrons, especially in metals and semiconductors. This separation suggests that inner electrons can be ignored in a large number of cases, thereby reducing the atom to an ionic core that interacts with the valence electrons. The use of an effective interaction, a pseudopotential, that approximates the potential felt by the valence electrons, was first proposed by Fermi in 1934 and Hellmann in 1935. In spite of the simplification pseudo-potentials introduce in calculations, they remained forgotten until the late 50's.

Ab Initio Pseudo-potentials

A crucial step toward more realistic pseudo-potentials was given by Topp and Hopfield and more recently Cronin, who suggested that the pseudo-potential should be adjusted such that they describe the valence charge density accurately. Based on that idea, modern pseudo-potentials are obtained inverting the free atom Schrödinger equation for a given reference electronic configuration and forcing the pseudo wave-functions to coincide with the true valence wave functions beyond a certain distance rl. The pseudo wave-functions are also forced to have the same norm as the true valence wave-functions and can be written as

$$R_l^{pp}(r) = R_{nl}^{AE}(r).$$

$$\int_0^{rl} dr \mid R_l^{PP}(r) \mid^2 r^2 = \int_0^{rl} dr \mid R_{nl}^{AE}(r) \mid^2 r^2.$$

where $R_l(r)$. l is the radial part of the wavefunction with angular momentum . The index n in the true wave-functions denotes the valence level. The distance beyond which the true and the pseudo wave-functions are equal, rl .

Electron Smearing

The electrons of system will occupy the lowest Kohn-Sham eigenstates up to a given energy level according to the Aufbau principle. This corresponds to the step-like Fermi-Dirac distribution at absolute zero. If there are several degenerate or close to degenerate eigenstates at the Fermi level, it is possible to get convergence problems, since very small perturbations may change the electron occupation. One way of damping these oscillations is to *smear* the electrons, i.e. allowing fractional occupancies. One approach of doing this is to assign a finite temperature to the electron Fermi-Dirac distribution. Other ways is to assign a cumulative Gaussian distribution of the electrons or using a Methfessel-Paxton method.

Software Supporting DFT

DFT is supported by many Quantum chemistry and solid state physics software packages, often along with other methods.

Time-dependent Density Functional Theory

Time-dependent density functional theory (TDDFT) is a quantum mechanical theory used in physics and chemistry to investigate the properties and dynamics of many-body systems in the presence of time-dependent potentials, such as electric or magnetic fields. The effect of such fields on molecules and solids can be studied with TDDFT to extract features like excitation energies, frequency-dependent response properties, and photoabsorption spectra.

TDDFT is an extension of density functional theory (DFT), and the conceptual and computational foundations are analogous – to show that the (time-dependent) wave function is equivalent to the (time-dependent) electronic density, and then to derive the effective potential of a fictitious non-interacting system which returns the same density as any given interacting system. The issue of constructing such a system is more complex for TDDFT, most notably because the time-dependent effective potential at any given instant depends on value of the density at all previous times. Consequently, the development of time-dependent approximations for the implementation of TDDFT is behind that of DFT, with applications routinely ignoring this memory requirement.

Overview

The formal foundation of TDDFT is the Runge-Gross (RG) theorem (1984) – the time-dependent analogue of the Hohenberg-Kohn (HK) theorem (1964). The RG theorem shows that, for a given initial wavefunction, there is a unique mapping between the time-dependent external potential of a system and its time-dependent density. This implies that the many-body wavefunction, depending upon $3N$ variables, is equivalent to the density, which depends upon only 3, and that all properties of a system can thus be determined from knowledge of the density alone. Unlike in DFT, there is no general minimization principle in time-dependent quantum mechanics. Consequently, the proof of the RG theorem is more involved than the HK theorem.

Given the RG theorem, the next step in developing a computationally useful method is to determine the fictitious non-interacting system which has the same density as the physical (interacting) system of interest. As in DFT, this is called the (time-dependent) Kohn-Sham system. This system is formally found as the stationary point of an action functional defined in the Keldysh formalism.

The most popular application of TDDFT is in the calculation of the energies of excited states of isolated systems and, less commonly, solids. Such calculations are based on the fact that the linear response function – that is, how the electron density changes when the external potential changes – has poles at the exact excitation energies of a system. Such calculations require, in addition to the exchange-correlation potential, the exchange-correlation kernel – the functional derivative of the exchange-correlation potential with respect to the density.

Formalism

Runge-Gross Theorem

The approach of Runge and Gross considers a single-component system in the presence of a time-dependent scalar field for which the Hamiltonian takes the form

$$\hat{H}(t) = \hat{T} + \hat{V}_{ext}(t) + \hat{W},$$

where T is the kinetic energy operator, W the electron-electron interaction, and $V_{ext}(t)$ the external potential which along with the number of electrons defines the system. Nominally, the external potential contains the electrons' interaction with the nuclei of the system. For non-trivial time-dependence, an additional explicitly time-dependent potential is present which can arise, for example, from a time-dependent electric or magnetic field. The many-body wavefunction evolves according to the time-dependent Schrödinger equation under a single initial condition,

$$\hat{H}(t)|\Psi(t)\rangle = i\hbar \frac{\partial}{\partial t}|\Psi(t)\rangle, \quad |\Psi(0)\rangle = |\Psi\rangle.$$

Employing the Schrödinger equation as its starting point, the Runge-Gross theorem shows that at any time, the density uniquely determines the external potential. This is done in two steps:

1. Assuming that the external potential can be expanded in a Taylor series about a given time, it is shown that two external potentials differing by more than an additive constant generate different current densities.

2. Employing the continuity equation, it is then shown that for finite systems, different current densities correspond to different electron densities.

Time-dependent Kohn-Sham System

For a given interaction potential, the RG theorem shows that the external potential uniquely determines the density. The Kohn-Sham approaches chooses a non-interacting system (that for which the interaction potential is zero) in which to form the density that is equal to the interacting system. The advantage of doing so lies in the ease in which non-interacting systems can be solved – the wave function of a non-interacting system can be represented as a Slater determinant of single-particle orbitals, each of which are determined by a single partial differential equation in three variable – and that the kinetic energy of a non-interacting system can be expressed exactly in terms of those orbitals. The problem is thus to determine a potential, denoted as $v_s(\mathbf{r},t)$ or $v_{KS}(\mathbf{r},t)$, that determines a non-interacting Hamiltonian, H_s,

$$\hat{H}_s(t) = \hat{T} + \hat{V}_s(t),$$

which in turn determines a determinantal wave function

$$\hat{H}_s(t)|\Phi(t)\rangle = i\frac{\partial}{\partial t}|\Phi(t)\rangle, \quad |\Phi(0)\rangle = |\Phi\rangle,$$

which is constructed in terms of a set of N orbitals which obey the equation,

$$\left(-\frac{1}{2}\nabla^2 + v_s(\mathbf{r},t)\right)\phi_i(\mathbf{r},t) = i\frac{\partial}{\partial t}\phi_i(\mathbf{r},t) \quad \phi_i(\mathbf{r},0) = \phi_i(\mathbf{r}),$$

and generate a time-dependent density

$$\rho_s(\mathbf{r},t) = \sum_{i=1}^{N} |\phi_i(\mathbf{r},t)|^2,$$

such that ρ_s is equal to the density of the interacting system at all times:

$$\rho_s(\mathbf{r},t) = \rho(\mathbf{r},t).$$

If the potential $v_s(\mathbf{r},t)$ can be determined, or at the least well-approximated, then the original Schrödinger equation, a single partial differential equation in $3N$ variables, has been replaced by N differential equations in 3 dimensions, each differing only in the initial condition.

The problem of determining approximations to the Kohn-Sham potential is challenging. Analogously to DFT, the time-dependent KS potential is decomposed to extract the external potential of the system and the time-dependent Coulomb interaction, v_J. The remaining component is the exchange-correlation potential:

$$v_s(\mathbf{r},t) = v_{\text{ext}}(\mathbf{r},t) + v_J(\mathbf{r},t) + v_{\text{xc}}(\mathbf{r},t).$$

In their seminal paper, Runge and Gross approached the definition of the KS potential through an action-based argument starting from the Dirac action

$$A[\Psi] = \int dt \, \langle \Psi(t) | H - i\frac{\partial}{\partial t} | \Psi(t) \rangle.$$

Treated as a functional of the wave function, $A[\Psi]$, variations of the wave function yield the many-body Schrödinger equation as the stationary point. Given the unique mapping between densities and wave function, Runge and Gross then treated the Dirac action as a density functional,

$$A[\rho] = A[\Psi[\rho]],$$

and derived a formal expression for the exchange-correlation component of the action, which determines the exchange-correlation potential by functional differentiation. Later it was observed that an approach based on the Dirac action yields paradoxical conclusions when considering the causality of the response functions it generates. The density response function, the functional derivative of the density with respect to the external potential, should be causal: a change in the potential at a given time can not affect the density at earlier times. The response functions from the Dirac action however are symmetric in time so lack the required causal structure. An approach which does not suffer from this issue was later introduced through an action based on the Keldysh formalism of complex-time path integration. An alternative resolution of the causality paradox through a refinement of the action principle *in real time* has been recently proposed by Vignale.

Linear Response TDDFT

Linear-response TDDFT can be used if the external perturbation is small in the sense that it does not completely destroy the ground-state structure of the system. In this case one can analyze the linear response of the system. This is a great advantage as, to first order, the variation of the system will depend only on the ground-state wave-function so that we can simply use all the properties of DFT.

Consider a small time-dependent external perturbation $\delta V^{ext}(t)$.. This gives

$$H'(t) = H + \delta V^{ext}(t)$$

$$H'_{KS}[\rho](t) = H_{KS}[\rho] + \delta V_H[\rho](t) + \delta V_{xc}[\rho](t) + \delta V^{ext}(t)$$

and looking at the linear response of the density

$$\delta\rho(\mathbf{r}t) = \chi(\mathbf{r}t,\mathbf{r}'t')\delta V^{ext}(\mathbf{r}'t')$$

$$\delta\rho(\mathbf{r}t) = \chi_{KS}(\mathbf{r}t,\mathbf{r}'t')\delta V^{eff}[\rho](\mathbf{r}'t')$$

where $\delta V^{eff}[\rho](t) = \delta V^{ext}(t) + \delta V_H[\rho](t) + \delta V_{xc}[\rho](t)$ Here and in the following it is assumed that primed variables are integrated.

Within the linear-response domain, the variation of the Hartree (H) and the exchange-correlation (xc) potential to linear order may be expanded with respect to the density variation

$$\delta V_H[\rho](\mathbf{r}) = \frac{\delta V_H[\rho]}{\delta\rho}\delta\rho = \frac{1}{|\mathbf{r}-\mathbf{r}'|}\delta\rho(\mathbf{r}')$$

and

$$\delta V_{xc}[\rho](\mathbf{r}) = \frac{\delta V_{xc}[\rho]}{\delta\rho}\delta\rho = f_{xc}(\mathbf{r}t,\mathbf{r}'t')\delta\rho(\mathbf{r}')$$

Finally, inserting this relation in the response equation for the KS system and comparing the resultant equation with the response equation for the physical system yields the Dyson equation of TDDFT:

$$\chi(\mathbf{r}_1 t_1,\mathbf{r}_2 t_2) = \chi_{KS}(\mathbf{r}_1 t_1,\mathbf{r}_2 t_2) + \chi_{KS}(\mathbf{r}_1 t_1,\mathbf{r}_2' t_{2'})\left(\frac{1}{|\mathbf{r}_{2'}-\mathbf{r}_{1'}|} + f_{xc}(\mathbf{r}_{2'} t_{2'},\mathbf{r}_{1'} t_{1'})\right)\chi(\mathbf{r}_{1'} t_{1'},\mathbf{r}_2 t_2)$$

From this last equation it is possible to derive the excitation energies of the system, as these are simply the poles of the response function.

Other linear-response approaches include the Casida formalism (an expansion in electron-hole pairs) and the Sternheimer equation (density-functional perturbation theory).

Lieb–Oxford Inequality

In quantum chemistry and physics, the Lieb–Oxford inequality provides a lower bound for the indirect part of the Coulomb energy of a quantum mechanical system. It is named after Elliott H. Lieb and Stephen Oxford.

The inequality is of importance for density functional theory and plays a role in the proof of stability of matter.

Introduction

In classical physics, one can calculate the Coulomb energy of a configuration of charged particles in the following way. First, calculate the charge density ρ, where ρ is a function of the coordinates $x \in \mathbb{R}^3$. Second, calculate the Coulomb energy by integrating:

$$\frac{1}{2}\int_{\mathbb{R}^3}\int_{\mathbb{R}^3}\frac{\rho(x)\rho(y)}{|x-y|}d^3x\,d^3y.$$

In other words, for each pair of points x and y, this expression calculates the energy related to the fact that the charge at x is attracted to or repelled from the charge at y. The factor of $\frac{1}{2}$ corrects for double-counting the pairs of points.

In quantum mechanics, it is *also* possible to calculate a charge density ρ, which is a function of $x \in \mathbb{R}^3$. More specifically, ρ is defined as the expectation value of charge density at each point. But in this case, the above formula for Coulomb energy is not correct, due to exchange and correlation effects. The above, classical formula for Coulomb energy is then called the "direct" part of Coulomb energy. To get the *actual* Coulomb energy, it is necessary to add a correction term, called the "indirect" part of Coulomb energy. The Lieb–Oxford inequality concerns this indirect part. It is relevant in density functional theory, where the expectation value ρ plays a central role.

Statement of the Inequality

For a quantum mechanical system of N particles, each with charge e, the N-particle density is denoted by

$$P(x_1,\ldots,x_N).$$

The function P is only assumed to be non-negative and normalized. Thus the following applies to particles with any "statistics". For example, if the system is described by a normalised square integrable N-particle wave function

$$\psi \in L^2(\mathbb{R}^{3N}),$$

then

$$P(x_1,\ldots,x_N) = |\psi(x_1,\ldots,x_N)|^2 .$$

More generally, in the case of particles with spin having q spin states per particle and with corresponding wave function

$$\psi(x_1,\sigma_1,\ldots,x_N,\sigma_N)$$

the N-particle density is given by

$$P(x_1,\ldots,x_N) = \sum_{\sigma_1=1}^{q} \cdots \sum_{\sigma_N=1}^{q} |\psi(x_1,\sigma_1,\ldots,x_N,\sigma_N)|^2 .$$

Alternatively, if the system is described by a density matrix γ, then P is the diagonal

$$\gamma(x_1,\ldots,x_N;x_1,\ldots,x_N).$$

The electrostatic energy of the system is defined as

$$I_P \quad e^2 \sum_{\le i < j \le N} \int_{\mathbb{R}} \frac{P(x,\ldots,x_i,\ldots,x_j,\ldots,x_N)}{|x_i \quad x_j|} d^3x \quad d^3x_N.$$

For $x \in \mathbb{R}^3$, the single particle charge density is given by

$$\rho(x) = |e| \sum_{i=1}^{N} \int_{\mathbb{R}^{3(N-1)}} P(x_1,\ldots,x_{i-1},x,x_{i+1},\ldots,x_N) d^3x_1 \cdots d^3x_{i-1} d^3x_{i+1} \cdots d^3x_N$$

and the direct part of the Coulomb energy of the system of N particles is defined as the electrostatic energy associated with the charge density ρ, i.e.

$$D(\rho) = \frac{1}{2} \int_{\mathbb{R}^3} \int_{\mathbb{R}^3} \frac{\rho(x)\rho(y)}{|x-y|} d^3x d^3y.$$

The Lieb–Oxford inequality states that the difference between the true energy I_P and its semiclassical approximation $D(\rho)$ is bounded from below as

$$E_P = I_P - D(\rho) \ge -C|e|^{\frac{2}{3}} \int_{\mathbb{R}^3} |\rho(x)|^{\frac{4}{3}} d^3x, \qquad (1)$$

where $C \le 1.68$ is a constant independent of the particle number N. E_P is referred to as the indirect part of the Coulomb energy and in density functional theory more commonly as the exchange plus correlation energy. A similar bound exists if the particles have different charges $e_1, \ldots, e_N$. No upper bound is possible for E_P.

The Optimal Constant

While the original proof yielded the constant $C = 8.52$, Lieb and Oxford managed to refine this result to $C = 1.68$. Later, the same method of proof was used to further improve the constant to $C = 1.64$. With these constants the inequality holds for any particle number N.

The constant can be further improved if the particle number N is restricted. In the case of a single particle $N = 1$ the Coulomb energy vanishes, $I_P = 0$, and the smallest possible constant can be computed explicitly as $C_1 = 1.092$. The corresponding variational equation for the optimal ρ is the Lane–Emden equation of order 3. For two particles ($N = 2$) it is known that the smallest possible constant satisfies $C_2 \ge 1.234$. In general it can be proved that the optimal constants C_N increase with the number of particles, i.e $C_N \le C_{N+1}$. Any lower bound on the optimal constant for fixed particle number N is also a lower bound on the optimal constant in (1) for arbitrary particle number. The largest presently known numerically obtained lower bound on C was proved for $N = 60$ where $C_{60} \geq 1.41$. This bound has been obtained by considering an exponential density. For the same particle number a uniform density gives $C_{60} \geq 1.34$. Contrary to previous believes, these results suggest that a uniform density is not the most challenging one for setting the lower bound to C. To summarise, the best known bounds for C are $1.41 \leq C \leq 1.64$.

The Dirac Constant

Historically, the first approximation of the indirect part E_p of the Coulomb energy in terms of the single particle charge density was given by Paul Dirac in 1930 for fermions. The wave function under consideration is

$$\psi(x_1, \sigma_1, \ldots, x_N, \sigma_N) = \frac{\det(\varphi_i(x_j, \sigma_j))}{\sqrt{N!}}.$$

With the aim of evoking perturbation theory, one considers the eigenfunctions of the Laplacian in a large cubic box of volume $|\Lambda|$ and sets

$$\varphi_{\alpha,k}(x, \sigma) = \frac{\chi_\alpha(\sigma)e^{2\pi i k \cdot x}}{\sqrt{|\Lambda|}},$$

where $\chi_1, \ldots, \chi_q$ forms an orthonormal basis of C^q. The allowed values of $k \in R^3$ are $n/|\Lambda|^{\nu}_3$ with $n \in Z^3$

+. For large N, $|\Lambda|$, and fixed $\rho = N|e|/|\Lambda|$, the indirect part of the Coulomb energy can be computed to be

$$E_p(\text{Dirac}) = -C|e|^{2/3} q^{-1/3} \rho^{4/3} |\Lambda|,$$

with $C = 0.93$.

This result can be compared to the lower bound (1). In contrast to Dirac's approximation the Lieb–Oxford inequality does not include the number q of spin states on the right-hand side. The dependence on q in Dirac's formula is a consequence of his specific choice of wave functions and not a general feature.

Generalisations

The constant C in (1) can be made smaller at the price of adding another term to the right-hand side. By including a term that involves the gradient of a power of the single particle charge density ρ, the constant C can be improved to 1.45. Thus, for a uniform density system $C \leq 1.45$.

Minnesota Functionals

Minnesota Functionals (Myz) are a group of approximate exchange-correlation energy functionals in density functional theory (DFT). They are developed by the group of Prof. Donald Truhlar at the University of Minnesota. These functionals are based on the meta-GGA approximation, i.e. they include terms that depend on the kinetic energy density, and are all based on complicated functional forms parametrized on high-quality benchmark databases. These functionals can be used for traditional quantum chemistry and solid-state physics calculations. The Myz functionals are widely used and tested in the quantum chemistry community. Independent evaluations of the strenghs and limitations of the Myz functionals with respect to various chemical properties have, however, cast doubts on the accuracy Minnesota functionals, with the newer functionals being less accurate than the older ones. Minnesota functionals are available in a large number of popular quantum chemistry computer programs.

Family of Functionals

Minnesota 05

The first family of Minnesota functionals, published in 2005, is composed by:

- M05: Global hybrid functional with 28% HF exchange.

- M05-2X Global hybrid functional with 56% HF exchange.

Minnesota 06

The '06 family represent a general improvement over the 05 family and is composed of:

- M06-L: Local functional, 0% HF exchange. Intended to be fast, good for transition metals, inorganic and organometallics.

- M06: Global hybrid functional with 27% HF exchange. Intended for main group thermochemistry and non-covalent interactions, transition metal thermochemistry and organometallics. It is usually the most versatile of the 06 functionals, and because of this large applicability it can be slightly worse than M06-2X for specific properties that require high percentage of HF exchange, such as thermochemistry and kinetics.

- M06-2X: Global hybrid functional with 54% HF exchange. It is the top performer within the 06 functionals for main group thermochemistry, kinetics and non-covalent interactions, however it cannot be used for cases where multi-reference species are or might be involved, such as transition metal thermochemistry and organometallics.

- M06-HF: Global hybrid functional with 100% HF exchange. Intended for charge transfer TD-DFT and systems where self-interaction is pathological.

Minnesota 08

The '08 family was created with the primary intent to improve the M06-2X functional form, retaining the performances for main group thermochemistry, kinetics and non-covalent interactions. This family is composed by two functionals with a high percentage of HF exchange, with performances similar to those of M06-2X:

- M08-HX: Global hybrid functional with 52.23% HF exchange. Intended for main group thermochemistry, kinetics and non-covalent interactions.

- M08-SO: Global hybrid functional with 56.79% HF exchange. Intended for main group thermochemistry, kinetics and non-covalent interactions.

Minnesota 11

The '11 family introduces range-separation in the Minnesota functionals and modifications in the functional form and in the training databases. These modifications also cut the number of functionals in a complete family from 4 (M06-L, M06, M06-2X and M06-HF) to just 2:

- M11-L: Local functional (0% HF exchange) with dual-range DFT exchange. Intended to be fast, to be good for transition metals, inorganic, organometallics and non-covalent interactions, and to improve much over M06-L.

- M11: Range-separated hybrid functional with 42.8% HF exchange in the short-range and 100% in the long-range. Intended for main group thermochemistry, kinetics and non-covalent interactions, with an intended performance comparable to that of M06-2X, and for TD-DFT applications, with an intended performance comparable to M06-HF.

Minnesota 12

The 12 family uses a Nonseparable (MN) functional form aiming to provide balanced performance for both chemistry and solid-state physics applications. It is composed by:

- MN12-L: A local functional, 0% HF exchange. The aim of the functional was to be very versatile and provide good computational performance and accuracy for energetic and structural problems in both chemistry and solid-state physics.

- MN12-SX: Screened-exchange (SX) hybrid functional with 25% HF exchange in the short-range and 0% HF exchange in the long-range. MN12-L was intended to be very versatile and provide good performance for energetic and structural problems in both chemistry and solid-state physics, at a computational cost that is intermediate between local and global hybrid functionals.

Runge–Gross Theorem

In quantum mechanics, specifically time-dependent density functional theory, the Runge–Gross theorem (RG theorem) shows that for a many-body system evolving from a given initial wavefunction, there exists a one-to-one mapping between the potential (or potentials) in which the system evolves and the density (or densities) of the system. The potentials under which the theorem holds are defined up to an additive purely time-dependent function: such functions only change the phase of the wavefunction and leave the density invariant. Most often the RG theorem is applied to molecular systems where the electronic density, $\rho(r,t)$ changes in response to an external scalar potential, $v(r,t)$, such as a time-varying electric field.

The Runge–Gross theorem provides the formal foundation of time-dependent density functional theory. It shows that the density can be used as the fundamental variable in describing quantum many-body systems in place of the wavefunction, and that all properties of the system are functionals of the density.

The theorem was published by Erich Runge and Eberhard K. U. Gross in 1984. As of January 2011, the original paper has been cited over 1,700 times.

Overview

The Runge–Gross theorem was originally derived for electrons moving in a scalar external field. Given such a field denoted by v and the number of electron, N, which together determine a Hamiltonian H_v, and an initial condition on the wavefunction $\Psi(t = t_0) = \Psi_0$, the evolution of the wavefunction is determined by the Schrödinger equation

$$\hat{H}_v(t)\,|\,\Psi(t)\rangle = i\frac{\partial}{\partial t}\,|\,\Psi(t)\rangle.$$

At any given time, the N-electron wavefunction, which depends upon $3N$ spatial and N spin coordinates, determines the electronic density through integration as

$$\rho(\mathbf{r},t) = N\sum_{s_1}\cdots\sum_{s_N}\int \mathrm{d}\mathbf{r}_2\cdots\int \mathrm{d}\mathbf{r}_N\,|\,\Psi(\mathbf{r}_1,s_1,\mathbf{r}_2,s_2,...,\mathbf{r}_N,s_N,t)\,|^2 .$$

Two external potentials differing only by an additive time-dependent, spatially independent, function, $c(t)$, give rise to wavefunctions differing only by a phase factor $\exp(-ic(t))$, and therefore the same electronic density. These constructions provide a mapping from an external potential to the electronic density:

$$v(\mathbf{r},t) + c(t) \rightarrow e^{-ic(t)}\,|\,\Psi(t)\rangle \rightarrow \rho(\mathbf{r},t).$$

The Runge–Gross theorem shows that this mapping is invertible, modulo $c(t)$. Equivalently, that the density is a functional of the external potential and of the initial wavefunction on the space of potentials differing by more than the addition of $c(t)$:

$$\rho(\mathbf{r},t) = \rho[v,\Psi_0](\mathbf{r},t) \leftrightarrow v(\mathbf{r},t) = v[\rho,\Psi_0](\mathbf{r},t)$$

Proof

Given two scalar potentials denoted as $v(\mathbf{r},t)$ and $v'(\mathbf{r},t)$, which differ by more than an additive purely time-dependent term, the proof follows by showing that the density corresponding to each of the two scalar potentials, obtained by solving the Schrödinger equation, differ.

The proof relies heavily on the assumption that the external potential can be expanded in a Taylor series about the initial time. The proof also assumes that the density vanishes at infinity, making it valid only for finite systems.

The Runge–Gross proof first shows that there is a one-to-one mapping between external potentials and current densities by invoking the Heisenberg equation of motion for the current density so as to relate time-derivatives of the current density to spatial derivatives of the external potential. Given this result, the continuity equation is used in a second step to relate time-derivatives of the electronic density to time-derivatives of the external potential.

The assumption that the two potentials differ by more than an additive spatially independent term, and are expandable in a Taylor series, means that there exists an integer $k \geq 0$, such that

$$u_k(\mathbf{r}) \equiv \frac{\partial^k}{\partial t^k}\left(v(\mathbf{r},t)-v'(\mathbf{r},t)\right)\bigg|_{t=t_0}$$

is not constant in space. This condition is used throughout the argument.

Step 1

From the Heisenberg equation of motion, the time evolution of the current density, $j(\mathbf{r},t)$, under the external potential $v(\mathbf{r},t)$ which determines the Hamiltonian H_v, is

$$i\frac{\partial \mathbf{j}(\mathbf{r},t)}{\partial t} = \langle\Psi(t)|[\hat{\mathbf{j}}(\mathbf{r}),\hat{H}_v(t)]|\Psi(t)\rangle.$$

Introducing two potentials v and v', differing by more than an additive spatially constant term, and their corresponding current densities j and j', the Heisenberg equation implies

$$\left\{\left\{\frac{\partial}{\partial t}\left(\mathbf{j}(\mathbf{r},t)-\mathbf{j}'(\mathbf{r},t)\right)\right\}\right|_{t=t_0} = \langle\Psi(t_0)|[\hat{\mathbf{j}}(\mathbf{r}),\hat{H}_v(t_0)-\hat{H}_{v'}(t_0)]|\Psi(t_0)\rangle$$

$$= \langle\Psi(t_0)|[\hat{\mathbf{j}}(\mathbf{r}),\hat{V}(t_0)-\hat{V}'(t_0)]|\Psi(t_0)\rangle$$

$$= i\rho(\mathbf{r},t_0)\nabla\left(v(\mathbf{r},t_0)-v'(\mathbf{r},t_0)\right)$$

The final line shows that if the two scalar potentials differ at the initial time by more than a spatially independent function, then the current densities that the potentials generate will differ infinitesimally after t_0. If the two potentials do not differ at t_0, but $u_k(\mathbf{r}) \neq 0$ for some value of k, then repeated application of the Heisenberg equation shows that

$$i^{k+1} \frac{\partial^{k+1}}{\partial t^{k+1}} \left(\mathbf{j}(\mathbf{r},t) - \mathbf{j}'(\mathbf{r},t) \right) \Bigg|_{t=t_0} = i\rho(\mathbf{r},t)\nabla i^k \frac{\partial^k}{\partial t^k} \left(v(\mathbf{r},t) - v'(\mathbf{r},t) \right) \Bigg|_{t=t_0},$$

ensuring the current densities will differ from zero infinitesimally after t_0.

Step 2

The electronic density and current density are related by a continuity equation of the form

$$\frac{\partial \rho(\mathbf{r},t)}{\partial t} + \nabla \cdot \mathbf{j}(\mathbf{r},t) = 0.$$

Repeated application of the continuity equation to the difference of the densities ρ and ρ', and current densities j and j', yields

$$\frac{\partial^{k+2}}{\partial t^{k+2}} (\rho(\mathbf{r},t) - \rho'(\mathbf{r},t)) \Bigg|_{t=t_0} = -\nabla \cdot \frac{\partial^{k+1}}{\partial t^{k+1}} \left(\mathbf{j}(\mathbf{r},t) - \mathbf{j}'(\mathbf{r},t) \right) \Bigg|_{t=t_0}, = -\nabla \cdot \rho \; \mathbf{r} \; t \; \nabla$$

$$= -\nabla \cdot [\rho(\mathbf{r},t_0) \nabla \frac{\partial^k}{\partial t^k} \left(v(\mathbf{r},t_0) - v'(\mathbf{r},t_0) \right) \Bigg|_{t=t_0}],$$

$$= -\nabla \cdot [\rho(\mathbf{r},t_0) \nabla u_k(\mathbf{r})].$$

The two densities will then differ if the right-hand side (RHS) is non-zero for some value of k. The non-vanishing of the RHS follows by a reductio ad absurdum argument. Assuming, contrary to our desired outcome, that

$$\nabla \cdot (\rho(\mathbf{r},t_0) \nabla u_k(\mathbf{r})) = 0,$$

integrate over all space and apply Green's theorem.

$$0 = \int d\mathbf{r} \; u_k(\mathbf{r}) \nabla \cdot (\rho(\mathbf{r},t_0) \nabla u_k(\mathbf{r})),$$

$$= -\int d\mathbf{r} \; \rho(\mathbf{r},t_0) (\nabla u_k(\mathbf{r}))^2 + \frac{1}{2} \int d\mathbf{S} \cdot \rho(\mathbf{r},t_0) (\nabla u_k^2(\mathbf{r})).$$

The second term is a surface integral over an infinite sphere. Assuming that the density is zero at infinity (in finite systems, the density decays to zero exponentially) and that $\nabla u_k^2(\mathbf{r})$ increases slower than the density decays, the surface integral vanishes and, because of the non-negativity of the density,

$$\rho(\mathbf{r}, t_0)(\nabla u_k(\mathbf{r}))^2 = 0,$$

implying that uk is a constant, contradicting the original assumption and completing the proof.

Extensions

The Runge–Gross proof is valid for pure electronic states in the presence of a scalar field. The first extension of the RG theorem was to time-dependent ensembles, which employed the Liouville equation to relate the Hamiltonian and density matrix. A proof of the RG theorem for multicomponent systems—where more than one type of particle is treated within the full quantum theory—was introduced in 1986. Incorporation of magnetic effects requires the introduction of a vector potential (A(r)) which together with the scalar potential uniquely determine the current density. Time-dependent density functional theories of superconductivity were introduced in 1994 and 1995. Here, scalar, vector, and pairing ($D(t)$) potentials map between current and anomalous ($\Delta_{IP}(r,t)$) densities.

Orbital-free Density Functional Theory

In computational chemistry, orbital-free density functional theory is a quantum mechanical approach to electronic structure determination which is based on functionals of the electronic density. It is most closely related to the Thomas–Fermi model. Orbital-free density functional theory is, at present, less accurate than Kohn–Sham density functional theory models, but has the advantage of being fast, so that it can be applied to large systems.

The Kinetic Energy of Electrons

The Hohenberg-Kohn theorems guarantee that, for a system of atoms, there exists a functional of the electron density that yields the total energy. Minimization of this functional with respect to the density gives the ground-state density from which all of the system's properties can be obtained. Although the Hohenberg-Kohn theorems tell us that such a functional exists, they do not give us guidance on how to find it. In practice, the density functional is known exactly except for two terms. These are the electronic kinetic energy and the exchange-correlation energy. The lack of the true exchange-correlation functional is a well known problem in DFT and there exists a huge variety of approaches to approximate this crucial component.

The fact that there is no known density functional for the electron kinetic energy is generally circumvented in another way. The traditional approach of density functional theory is to assume that the system can be treated as electrons residing in single-particle states (called orbitals). The total wave function can then be written as a Slater determinant of these single-particle orbitals. The orbitals themselves are found by diagonalizing the effective Kohn-Sham Hamiltonian. The kinetic energy of an electron in a single-particle state can be written exactly in terms of the orbital.

$$E_{kinetic} = -\frac{1}{2}\langle\phi_i|\nabla^2|\phi_i\rangle.$$

The problem with this approach is that it requires diagonalization of the Kohn-Sham Hamiltonian to find the single-particle orbitals. Further, since the Hamiltonian itself depends on these orbitals, the problem must be solved self-consistently. This is, in general, a computationally expensive process. If one could write the electron kinetic energy as a density functional, the problem of diagonalizing a large matrix could be replaced with a relatively straightforward functional optimization problem. Thus, finding an accurate density functional for the kinetic energy is the key focus of the so-called "orbital free" methods.

One of the first attempts to do this was the Thomas–Fermi model, which wrote the kinetic energy as

$$E_{TF} = \frac{3}{10}\left(3\pi^2\right)^{\frac{2}{3}}\int\left[n(\vec{r})\right]^{\frac{5}{3}}d^3r..$$

This expression is based on the homogeneous electron gas and, thus, is not very accurate for most physical systems. Finding more accurate and transferable kinetic energy density functionals is the focus of ongoing research.

Local-density Approximation

Local-density approximations (LDA) are a class of approximations to the exchange–correlation (XC) energy functional in density functional theory (DFT) that depend solely upon the value of the electronic density at each point in space (and not, for example, derivatives of the density or the Kohn–Sham orbitals). Many approaches can yield local approximations to the XC energy. However, overwhelmingly successful local approximations are those that have been derived from the homogeneous electron gas (HEG) model. In this regard, LDA is generally synonymous with functionals based on the HEG approximation, which are then applied to realistic systems (molecules and solids).

In general, for a spin-unpolarized system, a local-density approximation for the exchange-correlation energy is written as

$$E_{xc}^{LDA}[\rho] = \int \rho(\mathbf{r})\epsilon_{xc}(\rho)\,\mathbf{dr}\,,$$

where ρ is the electronic density and ε_{xc} is the exchange-correlation energy per particle of a homogeneous electron gas of charge density ρ. The exchange-correlation energy is decomposed into exchange and correlation terms linearly,

$$E_{xc} = E_x + E_c \,,$$

so that separate expressions for E_x and E_c are sought. The exchange term takes on a simple analytic form for the HEG. Only limiting expressions for the correlation density are known exactly, leading to numerous different approximations for ε_c.

Local-density approximations are important in the construction of more sophisticated approximations to the exchange-correlation energy, such as generalized gradient approximations or hybrid functionals, as a desirable property of any approximate exchange-correlation functional is that it reproduce the exact results of the HEG for non-varying densities. As such, LDA's are often an explicit component of such functionals.

Applications

Local density approximations, as with Generalised Gradient Approximations (GGA) are employed extensively by solid state physicists in ab-initio DFT studies to interpret electronic and magnetic interactions in semiconductor materials including semiconducting oxides and Spintronics. The importance of these computational studies stems from the system complexities which bring about high sensitivity to synthesis parameters necessitating first-principles based analysis. The prediction of Fermi level and band structure in doped semiconducting oxides is often carried out using LDA incorporated into simulation packages such as CASTEP and DMol3. However an underestimation in Band gap values often associated with LDA and GGA approximations may lead to false predictions of impurity mediated conductivity and/or carrier mediated magnetism in such systems.

Homogeneous Electron Gas

Approximation for ε_{xc} depending only upon the density can be developed in numerous ways. The most successful approach is based on the homogeneous electron gas. This is constructed by placing N interacting electrons in to a volume, V, with a positive background charge keeping the system neutral. N and V are then taken to infinity in the manner that keeps the density ($\rho = N / V$) finite. This is a useful approximation as the total energy consists of contributions only from the kinetic energy and exchange-correlation energy, and that the wavefunction is expressible in terms of planewaves. In particular, for a constant density ρ, the exchange energy density is proportional to $\rho^{\frac{4}{3}}$.

Exchange Functional

The exchange-energy density of a HEG is known analytically. The LDA for exchange employs this expression under the approximation that the exchange-energy in a system where the density is not homogeneous, is obtained by applying the HEG results pointwise, yielding the expression

$$E_x^{LDA}[\rho] = -\frac{3}{4}\left(\frac{3}{\pi}\right)^{1/3} \int \rho(\mathbf{r})^{4/3} \, d\mathbf{r}.$$

Correlation Functional

Analytic expressions for the correlation energy of the HEG are available in the high- and low-density limits corresponding to infinitely-weak and infinitely-strong correlation. For a HEG with density ρ, the high-density limit of the correlation energy density is

$$\epsilon_c = A\ln(r_s) + B + r_s(C\ln(r_s) + D),$$

and the low limit

$$\epsilon_c = \frac{1}{2}\left(\frac{g_0}{r_s} + \frac{g_1}{r_s^{3/2}} + \dots\right),$$

where the Wigner-Seitz radius is related to the density as

$$\frac{4}{3}\pi r_s^3 = \frac{1}{\rho}.$$

The analytical expression for the full range of densities has been proposed based on the many-body perturbation theory. The error as compared to the near-exact quantum Monte Carlo simulation is on the order of milli-Hartree.

- Chachiyo's correlation functional:

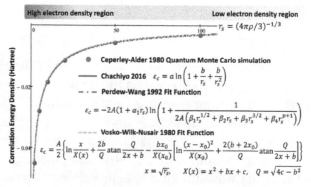

Comparison between several LDA correlation energy functionals and the quantum Monte Carlo simulation

$$\epsilon_c = a\ln\left(1 + \frac{b}{r_s} + \frac{b}{r_s^2}\right)$$

Accurate quantum Monte Carlo simulations for the energy of the HEG have been performed for several intermediate values of the density, in turn providing accurate values of the correlation energy density. The most popular LDA's to the correlation energy density interpolate these accurate values obtained from simulation while reproducing the exactly known limiting behavior. Various approaches, using different analytic forms for ε_c, have generated several LDA's for the correlation functional, including

- Vosko-Wilk-Nusair (VWN)

- Perdew-Zunger (PZ81)

- Cole-Perdew (CP)

- Perdew-Wang (PW92)

Predating these, and even the formal foundations of DFT itself, is the Wigner correlation functional obtained perturbatively from the HEG model.

Spin Polarization

The extension of density functionals to spin-polarized systems is straightforward for exchange, where the exact spin-scaling is known, but for correlation further approximations must be employed. A spin polarized system in DFT employs two spin-densities, ρ_α and ρ_β with $\rho = \rho_\alpha + \rho_\beta$, and the form of the local-spin-density approximation (LSDA) is

$$E_{xc}^{\text{LSDA}}[\rho_\alpha, \rho_\beta] = \int d\mathbf{r} \; \rho(\mathbf{r})\epsilon_{xc}(\rho_\alpha, \rho_\beta).$$

For the exchange energy, the exact result (not just for local density approximations) is known in terms of the spin-unpolarized functional:

$$E_x[\rho_\alpha, \rho_\beta] = \frac{1}{2}\left(E_x[2\rho_\alpha] + E_x[2\rho_\beta]\right).$$

The spin-dependence of the correlation energy density is approached by introducing the relative spin-polarization:

$$\zeta(\mathbf{r}) = \frac{\rho_\alpha(\mathbf{r}) - \rho_\beta(\mathbf{r})}{\rho_\alpha(\mathbf{r}) + \rho_\beta(\mathbf{r})}.$$

$\zeta = 0$ corresponds to the paramagnetic spin-unpolarized situation with equal α and β spin densities whereas $\zeta = \pm 1$ corresponds to the ferromagnetic situation where one spin density vanishes. The spin correlation energy density for a given values of the total density and relative polarization, $\varepsilon_c(\rho, \varsigma)$, is constructed so to interpolate the extreme values. Several forms have been developed in conjunction with LDA correlation functionals.

Exchange-correlation Potential

The exchange-correlation potential corresponding to the exchange-correlation energy for a local density approximation is given by

$$v_{xc}^{\text{LDA}}(\mathbf{r}) = \frac{\delta E^{\text{LDA}}}{\delta \rho(\mathbf{r})} = \epsilon_{xc}(\rho(\mathbf{r})) + \rho(\mathbf{r}) \frac{\partial \epsilon_{xc}(\rho(\mathbf{r}))}{\partial \rho(\mathbf{r})}.$$

In finite systems, the LDA potential decays asymptotically with an exponential form. This is in error; the true exchange-correlation potential decays much slower in a Coulombic manner. The artificially rapid decay manifests itself in the number of Kohn–Sham orbitals the potential can bind (that is, how many orbitals have energy less than zero). The LDA potential can not support a Rydberg series and those states it does bind are too high in energy. This results in the HOMO energy being too high in energy, so that any predictions for the ionization potential based on Koopman's theorem are poor. Further, the LDA provides a poor description of electron-rich species such as anions where it is often unable to bind an additional electron, erroneously predicating species to be unstable.

References

- Parr, Robert G.; Yang, Weitao (1989). Density-Functional Theory of Atoms and Molecules. New York: Oxford University Press. ISBN 0-19-509276-7.

- Fiolhais, Carlos; Nogueira, Fernando; Marques Miguel (2003). A Primer in Density Functional Theory. Springer. p. 60. ISBN 978-3-540-03083-6.

- Gross, E. K. U.; C. A. Ullrich; U. J. Gossman (1995). E. K. U. Gross and R. M. Dreizler, ed. Density Functional Theory. B. 337. New York: Plenum Press. ISBN 0-387-51993-9.

Permissions

Index

A

Adiabatic Chemical Dynamics, 4

Angular Momenta, 132, 181, 220-222, 231

Angular Momentum, 3, 27, 105-107, 113-114, 116, 121, 127, 131-132, 135, 138, 140, 142, 169-170, 173-178, 180-181, 183, 185, 201, 219-222, 226, 232, 234-236, 238, 241, 258

Asymptotic Behavior, 249

Atomic Orbital, 7-9, 12, 15, 84, 86, 127-129, 131, 134, 146-148, 153, 155-161, 164-165, 177, 180

Axial Symmetry, 226, 228-229, 232

B

Bohr Model, 3, 7, 9, 72, 102-106, 113-116, 121, 132, 134, 145, 167, 170, 173, 180

Bohr-sommerfeld Model, 113-115, 169

Born-oppenheimer Approximation, 1, 4, 6, 50-51, 54, 56, 58-59, 206, 252

C

Chaotic Systems, 195

Chemical Dynamics, 4

Complete Basis Set Methods (cbs), 69

Complex Orbitals, 135-136

Correlation Consistent Composite Approach (ccca), 69

Correlation Functional, 270, 272-274

Crystal Field Theory, 13, 72, 87, 93, 96, 100

Cycloadditions, 90

D

Density Functional Theory, 4, 15, 33, 62, 65, 67, 85, 206, 248-259, 261-263, 265, 267-271, 273, 275

Diatomic Molecule, 45, 47, 88, 156, 216-217, 219-221, 223, 225, 227-229, 231, 234-235, 238-239, 241, 243, 245, 247

Dirac Constant, 265

E

Electrocyclic Reactions, 92

Electron Configuration, 6-8, 10-13, 15-16, 47, 87, 128, 134, 143-144, 147, 158-162, 164

Electron Smearing, 258

Electronic Density, 4, 62, 65, 248-249, 258, 268-271

Electronic States, 47, 49-50, 54, 89, 218-219, 270

Electronic Structure, 2, 5, 9, 72-73, 77, 79, 81, 83, 85, 87, 89, 91, 93, 95-97, 99, 101, 103, 105, 107, 109, 111, 113, 115, 117, 121, 125, 131, 144, 148, 206-208, 214, 250, 252, 270

Exchange-correlation Potential, 254, 259, 261, 274-275

F

Feller-peterson-dixon Approach (fpd), 67

Franck-condon Principle, 46-54, 158

Frontier Molecular Orbital Theory, 72, 89-90

H

Heteronuclear Diatomics, 151, 155, 164

Heteronuclear Molecules, 217

Homogeneous Electron Gas, 61, 271-272

Hydrogen Atom, 2, 10, 12, 83, 102-103, 105-107, 109, 113, 115, 121, 125, 127, 133-135, 137, 140, 146, 154, 159, 167-171, 173-177, 196, 201, 209, 239

Hydrogen Electron Orbitals, 175

I

Interatomic Potential, 6, 28, 30, 32-33

Inversion Symmetry, 226, 229, 239

Isotropic Harmonic Oscillator, 42

J

Jellium, 6, 61-62, 64-65, 254

L

Lieb-oxford Inequality, 248, 262-263, 265

Ligand Field Theory, 17, 27, 72, 87, 93, 96-97

Local-density Approximation, 62, 65, 248, 251, 253, 271

M

Many-body Potentials, 30-31, 33

Minnesota Functionals, 248, 254, 265-267

Molecular Geometry, 66, 72-73, 78-79, 158, 218, 240-242

Molecular Orbital, 3, 9, 13, 15, 57, 72, 82-87, 89-90, 93, 95-96, 127, 130, 134, 146-153, 155-158, 161, 166, 214-215, 240, 242

Molecular Orbital Theory, 13, 15, 72, 83-87, 89-90, 93, 96, 127, 130, 147-148, 156, 215, 242

Molecular Symmetry, 6, 16-17, 19, 70, 94, 97, 166, 223-224

N

Non-adiabatic Chemical Dynamics, 4

Nuclear Cusp, 249

O

Odd-electron Molecules, 82

Old Quantum Theory, 10, 72, 103, 115-117, 124-126

Orbital Degeneracy, 152

Orbital-free Density Functional Theory, 248, 251, 270

P

Pair Potentials, 31

Periodic Table, 3, 7, 11-12, 111-112, 126, 128, 132, 134, 139, 143-145, 216

Perturbation Theory, 176, 188-189, 191-196, 199-203, 205-209, 262, 265, 273

Planetary Motion, 195

Potential Fitting, 33

Pseudo-potentials, 257

Q

Quantum Harmonic Oscillator, 6, 33, 39-40, 45-46, 221, 241

Quantum Mechanics, 1-3, 5-6, 8, 33-34, 40, 55, 70, 82, 85-86, 103, 106, 111, 113, 115, 117, 121, 125-129, 131-134, 136, 145, 168-169, 171, 173-174, 176-177, 187-188, 191, 194-196, 205, 209-210, 224-225, 235, 241, 246-248, 259, 263, 267

R

Real Orbitals, 136, 150

Repulsive Potentials, 32

Rotational Energies, 55, 220

Runge-gross Theorem, 248, 259-260, 267-268

S

Schoenflies Notation, 21, 228

Shell Model, 11, 110-111

Sigmatropic Reactions, 92

Singular Perturbation, 196, 200, 202-205

Spectrochemical Series, 95-98

Spin Polarization, 274

Spin-orbit Interaction, 180

Symmetry Operations, 19-21, 26-28, 224-227

T

Thermal Excitation, 241

Thomas-fermi Model, 4, 251, 256, 270-271

Time-dependent Density Functional Theory, 248, 251, 258, 267-268

Transition Metal Molecules, 81

Transition Metals, 11-13, 93, 214, 266-267

Translational Energies, 220

V

Valence Bond, 3, 72, 82-86, 88-89, 126, 209, 235-236, 242

Valence Bond Theory, 72, 82-86, 126, 209, 236

Valence Electrons, 9, 65, 67, 72-73, 85, 88, 145, 154, 167, 212-213, 251, 257

Variational Method, 188-189, 191, 193, 195, 199, 201, 203, 205, 207, 209, 239

Vibrational Energies, 220-221

Vibrational Energy Spacings, 221

W

Wave Model, 3

Z

Zero Differential Overlap, 211, 215

www.ingramcontent.com/pod-product-compliance
Lightning Source LLC
Jackson TN
JSHW052153130125
77033JS00004B/177